Lecture Notes in Computer Sci

T0238319

Commenced Publication in 1973
Founding and Former Series Editors:
Gerhard Goos, Juris Hartmanis, and Jan van Leeuwen

Jingshan Huang   Ryszard Kowalczyk
Zakaria Maamar   David Martin
Ingo Müller   Suzette Stoutenburg
Katia P. Sycara (Eds.)

Service-Oriented Computing: Agents, Semantics, and Engineering

AAMAS 2007 International Workshop, SOCASE 2007
Honolulu, HI, USA, May 14, 2007
Proceedings

 Springer

Volume Editors

Jingshan Huang
University of South Carolina, Columbia, SC 29208, USA
E-mail: huang27@sc.edu

Ryszard Kowalczyk
Swinburne University of Technology, Hawthorn, VIC 3122, Australia
E-mail: rkowalczyk@ict.swin.edu.au

Zakaria Maamar
Zayed University, PO Box 19282, Dubai, United Arab Emirates
E-mail: zakaria.maamar@zu.ac.ae

David Martin
SRI International, Menlo Park, CA 94025-3493, USA
E-mail: martin@ai.sri.com

Ingo Müller
Swinburne University of Technology, Hawthorn, VIC 3122, Victoria, Australia
E-mail: imueller@ict.swin.edu.au

Suzette Stoutenburg
The MITRE Corporation, Colorado Springs, Colorado 80910, USA
E-mail: suzette@mitre.org

Katia P. Sycara
Carnegie Mellon University, Pittsburgh, PA. 15213, USA
E-mail: katia@cs.cmu.edu

Library of Congress Control Number: 2007927059

CR Subject Classification (1998): H.3.5, H.3.3, H.3-4, I.2, C.2.4

LNCS Sublibrary: SL 3 – Information Systems and Application,
incl. Internet/Web and HCI

ISSN   0302-9743
ISBN-10   3-540-72618-7 Springer Berlin Heidelberg New York
ISBN-13   978-3-540-72618-0 Springer Berlin Heidelberg New York

Springer is a part of Springer Science+Business Media

springer.com

© Springer-Verlag Berlin Heidelberg 2007
Printed in Germany

Typesetting: Camera-ready by author, data conversion by Scientific Publishing Services, Chennai, India
Printed on acid-free paper   SPIN: 12066605   06/3180   5 4 3 2 1 0

Preface

The global trend towards more flexible and dynamic business process integration and automation has led to a convergence of interests between service-oriented computing, semantic technology, and intelligent multiagent systems. In particular the areas of service-oriented computing and semantic technology offer much interest to the multiagent system community, including similarities in system architectures and provision processes, powerful tools, and the focus on issues such as quality of service, security, and reliability. Similarly, techniques developed in the multiagent systems and semantic technology promise to have a strong impact on the fast-growing service-oriented computing technology.

Service-oriented computing has emerged as an established paradigm for distributed computing and e-business processing. It utilizes services as fundamental building blocks to enable the development of agile networks of collaborating business applications distributed within and across organizational boundaries. Services are self-contained, platform-independent software components that can be described, published, discovered, orchestrated, and deployed for the purpose of developing distributed applications across large heterogeneous networks such as the Internet.

Multiagent systems are also aimed at the development of distributed applications, however, from a different but complementary perspective. Service-oriented paradigms are mainly focused on syntactical and declarative definitions of software components, their interfaces, communication channels, and capabilities with the aim of creating interoperable and reliable infrastructures. In contrast, multiagent systems center on the development of reasoning and planning capabilities of autonomous problem solvers that apply behavioral concepts such as interaction, collaboration, or negotiation in order to create flexible and fault-tolerant distributed systems for dynamic and uncertain environments.

Semantic technology offers a semantic foundation for interactions among agents and services, forming the basis upon which machine-understandable service descriptions can be obtained, and as a result, autonomic coordination among agents is made possible. On the other hand, ontology-related technologies, ontology matching, learning, and automatic generation, etc., not only gain in potential power when used by agents, but also are meaningful only when adopted in real applications in areas such as service-oriented computing.

This volume consists of the proceedings of the Service-Oriented Computing: Agents, Semantics, and Engineering (SOCASE 2007) workshop held at the International Joint Conferences on Autonomous Agents and Multiagent Systems (AAMAS 2007). It also includes the four best papers selected from the Service-Oriented Computing and Agent-Based Engineering (SOCABE 2006) workshop held at AAMAS 2006. The papers in this volume cover a range of topics at the intersection of service-oriented computing, semantic technology, and intelligent multiagent systems, such as: service description and discovery; planning, composition and negotiation; semantic processes and service agents; and applications.

The workshop organizers would like to thank all members of the Program Committee for their excellent work, effort, and support in ensuring the high-quality program and successful outcome of the SOCASE 2007 workshop. We would also like to thank Springer for their cooperation and help in putting this volume together.

May 2007

Jingshan Huang
Ryszard Kowalczyk
Zakaria Maamar
David Martin
Ingo Müller
Suzette Stoutenburg
Katia Sycara

Organization

SOCASE 2007 was held in conjunction with The Sixth International Joint Conference on Autonomous Agents and Multiagent Systems (AAMAS 2007) on May 14, 2007 at the Hawaii Convention Center in Honolulu, Hawaii.

Organizing Committee

Jingshan Huang, University of South Carolina, USA
Ryszard Kowalczyk, Swinburne University of Technology, Australia
Zakaria Maamar, Zayed University Dubai, United Arab Emirates
David Martin, SRI International, USA
Ingo Müller, Swinburne University of Technology, Australia
Suzette Stoutenburg, The MITRE Corporation, USA
Katia Sycara, Carnegie Mellon University, USA

Program Committee

Esma Aimeur, University of Montreal, Canada
Stanislaw Ambroszkiewicz, Polish Academy of Sciences, Poland
Yacine Atif, United Arab Emirates University, United Arab Emirates
Youcef Baghdadi, Sultan Qaboos University, Oman
Djamal Benslimane, Université Claude Bernard Lyon 1, France
Jamal Bentahar, Concordia University Montreal, Canada
M. Brian Blake, Georgetown University, USA
Peter Braun, the agent factory GmbH, Germany
Paul A. Buhler, College of Charleston, USA
Bernard Burg, Panasonic Research, USA
Jiangbo Dang, Siemens Corporate Research, USA
Ian Dickinson, HP Laboratories Bristol, UK
Chirine Ghedira, Université Claude Bernard Lyon 1, France
Karthik Gomadam, University of Georgia, USA
Slimane Hammoudi, ESEO, France
Jingshan Huang, University of South Carolina, USA
Patrick Hung, University of Ontario, Canada
Nafaâ Jabeur, University of Windsor, Canada
Jugal Kalita, University of Colorado at Colorado Springs, USA
Mikko Laukkanen, TeliaSonera, Finland
Sandy Liu, NRC Institute for Information Technology, USA
Peter Mork, The MITRE Corporation, USA
Nanjangud C. Narendra, IBM India Research Lab, India
Manuel Núñez García, Universidad Complutense de Madrid, Spain

Table of Contents

Executing Semantic Web Services with a Context-Aware Service Execution Agent

António Luís Lopes and Luís Miguel Botelho

We, the Body, and the Mind Research Lab of ADETTI-ISCTE,
Avenida das Forças Armadas, Edifício ISCTE, 1600-082 Lisboa, Portugal
{antonio.lopes,luis.botelho}@we-b-mind.org

Abstract. The need to add semantic information to web-accessible services has created a growing research activity in this area. Standard initiatives such as OWL-S and WSDL enable the automation of discovery, composition and execution of semantic web services, i.e. they create a Semantic Web, such that computer programs or agents can implement an open, reliable, large-scale dynamic network of Web Services. This paper presents the research on agent technology development for context-aware execution of semantic web services, more specifically, the development of the Service Execution Agent (SEA). SEA uses context information to adapt the semantic web services execution process to a specific situation, thus improving its effectiveness and providing a faster and better service to its clients. Preliminary results show that context-awareness (e.g., the introduction of context information) in a service execution environment can speed up the execution process, in spite of the overhead that it is introduced by the agents' communication and processing of context information.

Keywords: Context-awareness, Semantic Web, Service Execution, Agents.

1 Introduction

Semantic Web Services are the driving technology of today's Internet as they can provide valuable information on-the-fly to users everywhere. Information-providing services, such as cinemas, hotels and restaurants information and a variety of e-commerce and business-to-business applications are implemented by web-accessible programs through databases, software sensors and even intelligent agents.

Research on Semantic Web standards, such as OWL-S [16] [19] and WSDL [3], opens the way for the creation of automatic processes for dealing with the discovery, composition and execution of web-based services. We have focused our research on the development of agent technology that allows the context-aware execution of semantic web services. We have decided to adopt the agent paradigm, creating SEA to facilitate the integration of this work in open agent societies [12], enabling these not only to execute semantic web services but also to seamlessly act as service providers in a large network of interoperating agents and web services.

J. Huang et al. (Eds.): SOCASE 2007, LNCS 4504, pp. 1–15, 2007.

In [21] the same approach was used in the Web Services infrastructure because of its capability to perform a range of coordination activities and mediation between requesters and providers. However, the broker-agent approach for the discovery and mediation of semantic web services in a multi-agent environment described in [21] does not take into account the use of context information. Thus we have decided to introduce context-awareness into the service execution process as a way of improving the services provided in dynamic multi-agent environments, such as the ones operating on pure peer-to-peer networks. Furthermore, the use of context information helps improve the execution process by adding valuable situation-aware information that will contribute to its effectiveness.

Being able to engage in complex interactions and to perform difficult tasks, agents are often seen as a vehicle to provide value-added services in open large-scale environments. However, the integration of agents as semantic web services providers is not easy due to the complex nature of agents' interactions. In order to overcome this limitation, we have decided to extend the OWL-S Grounding specification to enable the representation of services provided by intelligent agents. This extension is called the *AgentGrounding* and it is further detailed in [15].

We have also introduced the use of *Prolog* [4] for the formal representation of logical expressions in OWL-S control constructs. As far as we know, the only support for the formal representation of logical expressions in OWL-S (necessary for conditions, pre-conditions and effects) is done through the use of SWRL [13] and PDDL. Performance tests show that our *Prolog* approach improves the execution time of logical expressions in OWL-S services.

The remaining of this paper is organized as follows: section 2 gives a brief overview of related work; section 3 describes the use of context information and the introduction of context-aware capabilities in SEA; section 4 describes a motivating example which depicts a scenario where SEA is used; section 5 fully describes SEA by presenting its internal architecture, external interface, execution process and the implementation details; section 6 presents the performance tests and the overall evaluation of our research; finally, in section 7 we conclude the paper.

2   Related Work

The need to add semantic information to web-accessible services has created a growing research activity over the last years in this area. Regarding semantic web services, two major initiatives rise above the other, mainly because of their wide acceptance by the research community: WSMO [23] and OWL-S. A comparison between the two service description languages [14] concludes that the use of WSMO is more suitable for specific domains related to e-commerce and e-business, rather than for generic use, i.e., for different and more complex domains. OWL-S was designed to be generic and to cover a wide range of different domains but it lacked the formal specification to deal with logic expressions in the service description. We decided to use OWL-S as the Service Description Language due to its power in representing several different and complex domains. Other semantic web service execution approaches, such as WSMX [8] [9] are available but all rely on WSMO. However, since OWL-S was the chosen service description language to be used in this research, it is important to analyze the existing developed technology related to this standard in particular.

Two main software tools are referred in the OWL-S community as the most promising ones, regarding OWL-S services interpretation and execution: OWL-S VM [20] and OWL-S API [24]. However, at the time this research work has started, the OWL-S VM did not have a public release. OWL-S API is a developed Java API that supports the majority of the OWL-S specifications. For being the only OWL-S tool publicly available at the time, we have chosen to use and extend the OWL-S API.

In the interest of making the created technology interoperable with other systems that were already developed, we decided to ground its design and implementation on FIPA specifications, which are widely accepted agent standards. There are several existing FIPA-compliant systems that can be used: AAP [11], JADE [2], ZEUS [18], AgentAcademy [17] are just a few to be mentioned. We decided to use the JADE multi-agent platform because of its large community of users that constantly contribute to the improvement of the technology.

3   Service Execution and Context-Awareness

Context-aware computing is a computing paradigm in which applications can discover and take advantage of contextual information. As described in [7] "context is any information that can be used to characterize the situation of an entity, being an entity a person, place or object that is considered relevant to the interaction between a user and an application, including the user and applications themselves". We can enhance this definition of context by stating that the context of a certain entity is any information (provided by external sensors or other entities) that can be used to characterize the situation of that entity individually or when interacting with other entities. The same concept can be transferred to application-to-application interaction environments.

Context-aware computing can be summarized as being a mechanism that collects physical or emotional state information on a entity; analyses that information, either by treating it as an independent variable or by combining it with other information collected in the past or present; performs some action based on the analysis; and repeats from the first step, with some adaptation based on previous iterations [1].

SEA uses a similar approach as the one described in [1] to enhance its activity, by adapting it to the specific situation that the agent and its client are involved in, at the time of the execution process.

SEA interacts with a generic context system [5] in order to obtain context information, subscribe desired context events and to provide relevant context information. Other agents, web services and sensors (both software and hardware) in the environment will interact with the context system as well, by providing relevant context information related to their own activities, which may be useful to other entities in the environment.

Throughout the execution process, SEA provides and acquires context information from and to this context system. For example, SEA provides relevant context information about itself, such as its queue of service execution requests and the average time of service execution. This will allow other entities in the environment to determine the service execution agent with the smallest work load, and hence that can provide a faster execution service.

During the execution of a compound service, SEA invokes atomic services from specific service providers (both web services, and service provider agents). SEA also provides valuable information regarding these service providers' availability and average execution time to the context system. Other entities can use this information (by contacting the context system) to rate service providers or to simply determine the best service provider to use in a specific situation.

Furthermore, SEA uses its own context information (as well as information from other sources and entities in the environment) to adapt the execution process to a specific situation. For instance, when selecting among several providers of some service, SEA will choose the one with better availability (with less history of being offline) and lower average execution time.

In situations such as the one where service providers are unavailable, it is faster to obtain the context information from the context system (as long as service providers can also provide context information about their own availability) than by simply trying to use the services and finding out that they are unavailable (because of the time lost waiting for connection time-outs to occur). After obtaining this relevant information, SEA can then contact other service-oriented agents (such as service discovery and composition agents) for requesting the re-discovering of service providers and/or the re-planning of composed services. This situation-aware approach using context information on-the-fly helps SEA providing a value-added execution service.

4   Example: Search Books' Prices

In this section we present an example scenario in the domain of books' prices searching, in order to better prove the need for the existence of a broker agent for the execution of semantic web services.

Imagine a normal web user that wants to find a specific book (of which he doesn't recall the exact title) of a certain author, at the best price available. Probably, he would start by using a domain-specific search engine to find the intended item. After finding the exact book, he would then try to find websites that sell it. After doing an extensive search, he would finally find the web site that sells the book at the best price, but it only features the price in US dollars. The user is Portuguese and he would like to know the book's price in Euros, which leaves him with a new search for a currency converter service that would help him with this task. As we can see, this user would have to handle a lot of different web sites and specific searches to reach its objective: find the best price of a certain book.

The composition of atomic semantic web services into more complex value-added services would present an easy solution to this problem. The idea is to provide a unique compound service with the same features as the example described above, but the use of which would be a lot simpler, since it would be through the interaction with a service execution broker agent.

Fig. 1 shows the overall scenario description, by presenting all the participants and the interactions between them. We will assume the existence of a compound service called "book best-price search". The dashed gray lines represent the interactions that are not subject of this paper.

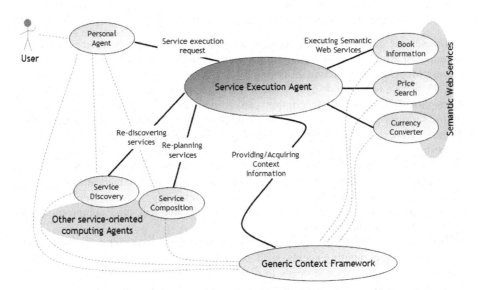

Fig. 1. Overall Scenario Description

The user of this service is represented in the figure as the client user and by his personal agent. The client user will provide the necessary information (author's name, book's title and expected currency) to his personal agent so that this can start the interaction with the remaining participants in order to obtain the desired result of the "book best-price search" service.

In order to request the execution of the "book best-price search" service, the personal agent needs to have the service's OWL-S description. This could have been previously obtained from a composition planner agent or service discovery agent. This interaction and the composition of the compound service are not covered by this paper.

After sending the "book best-price search" service's OWL-S description and the instantiation information (input parameters) provided by its user to the service execution agent, the personal agent will wait for the result of the service to then inform its user.

Semantic Web Services can be any web-based computer application, such as web services or intelligent agents, as long as they are described using a formal description language, such as OWL-S. They are represented in the figure on the opposite end to the client user. For this example, we'll consider the existence of the following semantic web services:

- *Book Information Agent* – this web-accessible information agent provides information about books and its authors
- *Price Search Web Services* – these web-accessible services provide an item's best-price search service (such as Amazon and Barnes & Noble)
- *Currency Converter Web Service* – this web service provides simple conversion of a monetary value from one currency to another.

The service execution agent interacts with these semantic web services to obtain the required information, according to instructions in the compound service's OWL-S Process Model and Grounding descriptions.

The service execution agent bases the execution on the service's OWL-S description. This OWL-S description can, sometimes, be incomplete, i.e., missing atomic services information regarding Grounding information. This can compromise the service's execution simply because the service execution agent doesn't have the necessary information to continue. On the other hand, if the service's OWL-S description is complete but the service execution agent is operating on very dynamic environments (such as pure P2P networks), the information contained in the service's OWL-S description can be easily out-dated (previously existing semantic web services are no longer available in the network). This will also compromise the agent's service execution activity.

To solve this problem, the service execution agent can interact with other service-oriented computing agents, such as service discovery and service composition agents, for example, when it needs to discover new services or when it needs to do some re-planning of compound services that somehow could not be executed, for whatever reasons explained above.

5   Service Execution Agent

The Service Execution Agent (SEA) is a broker agent that provides context-aware execution of semantic web services. The agent was designed and developed considering the interactions described in sections 3 and 4 and the internal architecture was clearly designed to enable the agent to receive requests from client agents, acquire/provide relevant context information, interacting with other service coordination agents when relevant and execute remote web services.

This section of the paper is divided into four sub-sections. Sub-sections 5.1 and 5.2 describe the internal architecture of the agent, explaining in detail the internal components and their interactions both internal and external, using the agent FIPA-ACL interface. Sub-section 5.3 describes the execution process that the agent carries out upon request from a client agent. Sub-section 5.4 provides some details on the implementation of the agent.

5.1   Internal Architecture

The developed agent is composed of three components: the Agent Interaction Component (AIC), the Engine Component (EC) and the Service Execution Component (SEC). Fig. 2 illustrates the internal architecture of the agent and the interactions that occur between the components and with external entities.

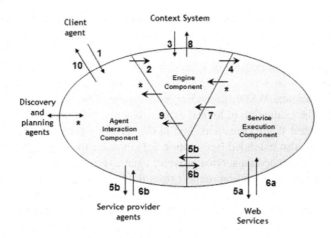

Fig. 2. SEA Internal Architecture and Interactions

The AIC was developed as an extension of the JADE platform and its goal is to provide an interaction framework to FIPA-compliant agents, such as SEA's clients (requesting the execution of specified services – Fig. 2, step 1) and service discovery and composition agents (when SEA is requesting the re-discovering and re-planning of specific services – Fig. 2, steps *). This component extends the JADE platform to provide extra features regarding language processing, behaviour execution, database information retrieval and components' communication. Among other things, the AIC is responsible for receiving messages, parsing them and processing them into a suitable format for the EC to use it (Fig. 2, step 2). The reverse process is also the responsibility of the AIC – receiving data from the EC and processing it into the agents' suitable format to be sent as messages Fig. 2, step 9).

The EC is the main component of SEA as it controls the agent's overall activity. It is responsible for pre-processing service execution requests, interacting with the context system and deciding when to interact with other agents (such as service discovery and composition agents). When the EC receives an OWL-S service execution request (Fig. 2, step 2), it acquires suitable context information (regarding potential service providers and other relevant information, such as client location – Fig. 2, step 3) and schedules the execution process. If the service providers of a certain atomic service (invoked in the received composed service) are not available, SEA interacts with a service discovery agent (through the AIC – Fig. 2, steps *) to discover available providers for the atomic services that are part of the OWL-S compound service. If the service discovery agent cannot find adequate service providers, the EC can interact with a service composition agent (again through the AIC – Fig. 2, steps *) asking it to create an OWL-S compound service that produces the same effects as the original service. After having a service ready for execution, with suitable context information, the EC sends it to the SEC (Fig. 2, step 4), for execution. Throughout the execution process, the EC is also responsible for providing context information to the context system, whether it is its own information (such as service execution requests' queue and average time of execution) or other entities' relevant context information (such as availability of providers and average execution time of services).

The SEC was developed as an extension of the OWL-S API and its goal is to execute semantic web services (Fig. 2, steps 5a and 6a) described using OWL-S service description and WSDL grounding information. The extension of the OWL-S API allows for the evaluation of logical expressions in conditioned constructs, such as the *If-then-Else* and *While* constructs, and in the service's pre-conditions and effects. OWL-S API was also extended (see section 5.3) in order to support the execution of services that are grounded on service provider agents (Fig. 2, steps 5b, 6b). When the SEC receives a service execution request from the EC, it executes it according to the description of the service's process model. This generic execution process is described in section 5.3. After execution of the specified service and generation of its results, the SEC sends them to the EC (Fig. 2, step 7) for further analysis and post-processing, which includes sending gathered context information to the context system (Fig. 2, step 8) and sending the results to the client agent (through the AIC – Fig. 2, steps 9, 10).

5.2 Agent Interface

When requesting the execution of a specified service, client agents interact with the Service Execution Agent through the FIPA-request interaction protocol [10]. This protocol states that when the receiver agent receives an action request, it can either agree or refuse to perform the action. In the first part of this protocol, the execution agent should then notify the other agent of its decision through the corresponding communicative act (FIPA-agree or FIPA-refuse). The execution agent performs this decision process through a service execution's request evaluation algorithm that involves acquiring adequate context information. The Service Execution Agent will only agree to perform a specific execution if it is possible to execute it, according to currently available context information. For example, if necessary service providers are not available and the time that requires finding alternatives is longer than the timeframe in which the client agent expects to obtain a reply to the execution request, then the execution agent refuses to perform it. On the other hand, if the execution agent is able to perform the execution request (because service providers are immediately available), but not in the time frame requested by the client agent (again, according to available context information) it also refuses to do so. The execution agent can also refuse to perform execution requests if its work load is already too high (if its requests queue is bigger than a certain threshold).

The FIPA-request also states that after successful execution of the requested action, the execution agent should return the corresponding results through a FIPA-inform message. After executing a service, SEA can send one of two different FIPA-inform messages: one sending the results obtained from the execution of the service; other sending just a notification that the service was successfully executed (when no results are produced by the execution).

5.3 Execution Process

OWL-S is an OWL-based service description language. OWL-S descriptions consist of three parts: a *Profile*, which tells "what the service does"; a *Process Model*, which tells "how the service works"; and a *Grounding*, which specifies "how to access a particular provider for the service". The Profile and Process Model are considered to be *abstract* specifications, in the sense that they do not specify the details of particular message formats, protocols, and network addresses by which a Web service is instantiated. The role of providing more concrete details belongs to the grounding part. WSDL (Web Service Description Language) provides a well-developed means of specifying these kinds of details. For the execution process, the most relevant parts of an OWL-S service description are the *Process Model* and the *Grounding*. The *Profile* part is more relevant for discovery, matchmaking and composition processes, hence no further details will be provided in this paper.

The *Process Model* describes the steps that should be performed for a successful service execution. These steps represent two different views of the process: first, a process produces a data transformation of the set of given inputs into the set of produced outputs; second, a process produces a transition in the world from one state to another. This transition is described by the preconditions and effects of the process [19].

The *Process Model* identifies three types of processes: *atomic*, *simple*, and *composite*. *Atomic* processes are directly evocable. They have no sub-processes, and can be executed in a single step, from the perspective of the service requester. *Simple* processes are not evocable and are not associated with a grounding description, but, like atomic processes, they *are* conceived of as having single-step executions. *Composite* processes are decomposable into other (non-composite or composite) processes. These represent several-steps executions, which can be described using different control constructs, such as *Sequence* (representing a sequence of steps) or *If-Then-Else* (representing conditioned steps).

The *Grounding* specifies the details of how to access the service. These details mainly include protocol and message formats, serialization, transport, and addresses of the service provider. The central function of an OWL-S Grounding is to show how the abstract inputs and outputs of an atomic process are to be concretely realized as messages, which carry those inputs and outputs in some specific format. The *Grounding* can be extended to represent specific communication capabilities, protocols or messages. WSDL and *AgentGrounding* are two possible extensions.

The general approach for the execution of OWL-S services consists of the following sequence of steps: (i) validate the service's pre-conditions, whereas the execution process continues only if all pre-conditions are true; (ii) decompose the compound service into individual atomic services, which in turn are executed by evoking their corresponding service providers using the description of the service providers contained in the grounding section of the service description; (iii) validate the service's effects by comparing them with the actual service execution results, whereas the execution process only proceeds if the service has produced the expected effects; (iv) collect the results, if any have been produced since the service may be only a "*change-the-world*" kind of service, and send them to the client who requested the execution.

WSDL is an XML format for describing network services as a set of endpoints operating on messages containing either document-oriented or procedure-oriented information. The operations and messages are described abstractly, and then bound to a concrete network protocol and message format to define an endpoint. Related concrete endpoints are combined into abstract endpoints (services). WSDL is extensible to allow description of endpoints and their messages regardless of what message formats or network protocols are used to communicate [3]. In short, WSDL describes the access to a specific service provider for a described OWL-S service.

However, WSDL currently lacks an efficient way of representing agent bindings, i.e., a representation for complex interactions such as the ones that take place with service provider agents. To overcome this limitation, we decided to create an extension of the OWL-S *Grounding* specification, named *AgentGrounding* [15]. This extension is the result of an analysis of the necessary requirements for interacting with agents when evoking the execution of atomic services. In order for an agent to act as a service provider in the Semantic Web, its complex communication schema has to be mapped into the OWL-S Grounding structure. To achieve this mapping, the *AgentGrounding* specification includes most of the elements that are present in Agent Communication Languages. At the message level, the *AgentGrounding* specification includes elements such as the name and the address of the service provider, the protocol and ontology that are used in the process, the agent communication language and the content language. At the message content level, the *AgentGrounding* specification includes elements such as name and type of the service to be evoked and its input and output arguments, including the types and the mapping to OWL-S parameters. For more details and a complete example, we refer the reader to [15].

In order to allow the specification of conditioned constructs which were based on logical expressions, we have introduced the use of *Prolog* for the formal representation of pre-conditions and *If* clauses in the descriptions of the services. Until now, the only support for control constructs that depend on formal representation of logical expressions in OWL-S was done through the use of SWRL. However, for performance reasons, which are shown in section 6, we have decided to use *Prolog*.

5.4 Implementation

The Service Execution Agent was implemented using Java and component-based software as well as other tools that were extended to incorporate new functionalities into the service execution environment. These tools are the JADE agent platform [2] and the OWL-S API [24].

The JADE agent platform was integrated into the Agent Interaction Component (see section 5.1) of the Service Execution Agent to enable its interaction with client agents and service provider agents.

The OWL-S API is a Java implementation of the OWL-S execution engine, which supports the specification of WSDL groundings. The OWL-S API was integrated into the Execution Component of the Service Execution Agent to enable it to execute both atomic and compound services.

In order to support the specification and execution of *AgentGrounding* descriptions (see section 5.3), we have extended the OWL-S API's execution process engine. The extension of the execution engine allows converting *AgentGrounding* descriptions into agent-friendly messages, namely FIPA-ACL messages, and sending these to the corresponding service provider agents.

To enable the support for control constructs that depend on formal representation of logical expressions using *Prolog*, we extended the OWL-S API with a *Prolog* engine that is able to process logical expressions present in *If/While* clauses and pre-conditions. This extension was done through the use of *TuProlog* [6], an open source Java-based *Prolog*.

6   Evaluation and Results

We have stated before that context information would be used to enhance the semantic web service execution process. The referred enhancement consists of the adaptation of the execution agent to a specific situation in a way such that the execution process is done according to a certain situation characteristics, as perceived from available context information. A situation is characterized by the properties that are imposed to the execution agent when this receives an execution request from a client. These properties can be simple limitations such as the client's request for the execution to be done in a certain time-frame or that the service has to be available near a specific location. In this section, we present the evaluation process that illustrates the advantage of using context information in the execution process carried out by the Service Execution Agent.

As described in section 3, the Service Execution Agent uses context information such as availability (to check if service providers often have been offline in the past), average execution time (the time each service provider takes to execute a single service, in average) and queues of pending requests (the number of requests waiting to be processed on each service provider) to determine who the "best" service providers are. This collected context information allows the execution agent to build a sort of *rating* schema of the available service providers, by applying the following simple formula (for each service provider): `Average Execution Time X Queue of Pending Requests + Average Execution Time = Estimated Time of Execution`. Combined with the availability history of the service provider (`Estimated Time of Execution X Number of Times Offline in the Last 10 Minutes`) this information will provide the rating (in reverse order) of the service provider within the list of available providers for the same service. Thus, when executing, the Service Execution Agent can use this information to determine the fastest way to execute a compound service or find alternatives in case of failure.

Even though this approach of adding context information to the execution process improves the way compound services are executed (by allowing the determination of the best service providers in a specific situation and by providing a failure recovery method), it introduces an *overhead* that doesn't exist in semantic web services execution environments that do not consider using context information. This *overhead* is composed of the procedures that the execution agent must perform when dealing with the remaining elements of the service coordination infrastructure: communicating and managing conversations with other agents (client, service discovery and composition agents), retrieving and processing context information (related to the service providers) and preparing the execution of compound services. It is important to determine how this *overhead* influences the overall execution time.

Table 1 shows the average behavior of the Service Execution Agent in the repeated execution (50 times) of the compound service (with 5 atomic services) described in section 4. The Normal Execution Time (NET) is the time that it takes to execute the web services; the Overhead Execution Time (OET) is the time that SEA takes to perform the mentioned procedures related to context information processing, conversations' management and preparation of execution; the Total Execution Time (TET) is the total time of execution: NET + OET.

Table 1. Execution time details (in seconds) of the repeated execution (50 requests) of the "book best-price search" service

	Total	Average	Max	Min
TET	1244,35	25,39	33,8	20,15
NET	1242,46	25,36	33,79	20,12
OET	1,89	0,04	0,15	0,01

Table 1 provides the execution time details of the entire test. As we can see in Table 1 the average overhead execution time is very small (0,04 seconds). This value is hardly noticeable by a user of such a system. Moreover, in some tests the use of this kind of service execution approach allowed the reduction of the overall execution time due to the fact that the execution agent always tried to find the fastest service provider (the one which was online for longer periods in the past or had a lower work load or had a lower average execution time).

It is important to detail the *overhead* execution time in order to determine which are the most time consuming processes. Fig. 3 shows the overall behavior of the *overhead* during the test described in Table 1.

The gray dashed line represents the time used in the communication with other agents; the regular black line represents the service processing time (gathering context

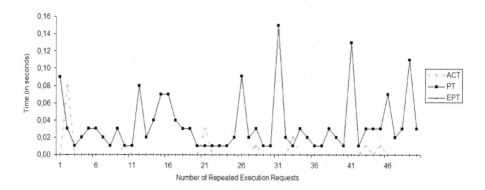

Fig. 3. Detailed *Overhead* Execution Time of the test described in Table 1

information and adapting accordingly); and the dashed black line (hardly noticeable in the figure because of being so low) represents the service execution process preparation, which mainly consists of changing the service description whenever necessary (for example, when service providers change and the grounding have to be changed accordingly). By analyzing the chart in Fig. 3, we can easily determine that the specific task that contributes to increase the *overhead* execution time is the preparation task, i.e., the access to the context system to retrieve relevant context information. We can also see that, in some cases, the time used in communication with other agents is the most time-consuming event. However, this can be explained by the environment in which agents operate, whereas the communication process in highly dynamic environments is influenced by network traffic. A deeper analysis of the preparation time allowed us to determine that the most time-consuming process when accessing the context system is in fact the communication with the context system through its JINI interface [5]. So, the preparation time can be reduced if the access to the context system can be improved through the use of a different communication interface, which proves to be faster and more reliable.

The described test was done using the "book best-price search" service description with logical expressions described in Prolog. However, to compare our approach of using Prolog with the existing approach originally implemented in the OWL-S API, we have decided to repeat the test, this time using SWRL for the representation of logical expressions. Fig. 4 details the execution times for each atomic service of the "book best-price search" service, this time using SWRL. The use of logical expressions is only done in the atomic service "Compare Prices", which consists on a *If* clause that compares the prices of both book shops (Amazon and Barnes & Noble) and determines which one is cheaper.

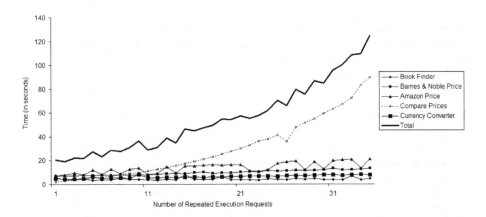

Fig. 4. Execution test of the "book best-price search" service with SWRL expressions

As we can see in Fig. 4, the "Compare Prices" atomic service is the only one that does not have a constant behavior throughout the entire execution test. In fact, this service increases its execution time on each cycle, thus contributing to the increase of the overall execution time of the compound service. This behavior may be explained by an inefficient implementation of the SWRL extension or the SPARQL [22] engine (used for the queries of logical expressions) in the OWL-S API, however we were not able to determine the exact cause of this inefficient behavior.

The overall evaluation of the work shows that the Service Execution Agent can be used in highly dynamic environments since it can efficiently distribute the *execution work* to the appropriate service providers, hence providing a faster and more reliable service to a user with time, device and location constraints.

7 Conclusions

We have presented a framework to enable the execution of semantic web services using a context-aware broker agent – the Service Execution Agent. The developed approach uses context information to determine the appropriate service providers for each situation. In doing so, SEA becomes a highly efficient broker agent capable of distributing the execution work among the available service providers, thus providing a more useful and faster service to its clients. Evaluation results show that the introduction of context-aware capabilities and the interaction within a service coordination infra-structure not only add a very small overhead to the execution process, as it can sometimes improve the overall execution time, thus offering a valid and reliable alternative to the existing semantic web services execution environments.

Future work will be based on the further analysis of the use of *Prolog* and SWRL in the logical representation of conditions, in order to determine the exact cause of the inefficient behaviour when using SWRL and whether *Prolog* is a better formal representation solution for logical expressions in the execution of semantic web services. Also, we intend to perform comparison tests of our approach with other semantic web services execution approaches that may have been recently developed.

Acknowledgments. This work has been supported in part by the European Commission under the project grant FP6-IST-511632-CASCOM. The authors would like to thank the reviewers for providing such useful comments for the overall improvement of the paper.

References

1. Abowd, G. D., Dey, A., Orr, R. and Brotherton, J. (1998). Context-awareness in wearable and ubiquitous computing. Virtual Reality, 3:200–211.
2. Bellifemine F., Poggi A., Rimassa G. (2001) Developing multi-agent systems with a FIPA-compliant agent framework, Software-Practice and Experience 31 (2): 103–128 Feb 2001
3. Christensen, E., Curbera, F., Meredith, G. and Weerawarana, S. (2001). Web Services Description Language (WSDL) 1.1. Available on-line at http://www.w3.org/TR/2001/NOTE-wsdl-20010315.
4. Clocksin, W.F.; Mellish, C.S. (1981). Programming in Prolog. Springer-Verlag, New York.
5. Costa, P., Botelho, L. (2005). Generic Context Acquisition and Management Framework. First European Young Researchers Workshop on Service Oriented Computing. Forthcoming.

6. Denti, E., Omicini, A., Ricci, A. (2005). Multi-paradigm Java-Prolog integration in TuProlog. Science of Computer Programming, Vol. 57, Issue 2, pp. 217-250.

7. Dey, A. K. and Abowd, G. D. (1999). Towards a better understanding of context and context awareness. GVU Technical Report GIT-GVU-99-22, College of Computing, Georgia Institute of Technology.

8. Domingue, J., Cabral, L., Hakimpour, F., Sell, D., and Motta, E. (2004). IRS-III: A Platform and Infrastructure for Creating WSMO-based Semantic Web Services. 3rd International Semantic Web Conference (ISWC2004), LNCS 3298.

9. Fensel, D. and Bussler, C. (2002). The Web Service Modeling Framework – WSMF. Electronic Commerce: Research and Applications, 1 (2002) 113-137.

10. FIPA Members. (2002) Foundation for Intelligent Physical Agents website. http://www.fipa.org/.

11. Fujitsu Labs of America. (2001). April Agent Platform project website. http://www.nar.fujitsulabs.com/aap/about.html

12. Helin, H., Klusch, M., Lopes, A., Fernández, A., Schumacher, M., Schuldt, H., Bergenti, F., Kinnunen, A. (2005). CASCOM: Context-aware Service Co-ordination in Mobile P2P Environments. Lecture Notes in Computer Science, Vol. 3550 / 2005, ISSN: 0302-9743, pp. 242 - 243

13. Horrocks, I., Patel-Schneider, P.F., Boley, H., Tabet, S., Grosof, B., Dean, M. (2004). SWRL: A Semantic Web Rule Language combining OWL and RuleML. W3C Member Submission, available on-line at http://www.w3.org/Submission/2004/SUBM-SWRL-20040521/

14. Lara, R., Roman, D., Polleres, A., Fensel, D. (2004). A Conceptual Comparison of WSMO and OWL-S. Proceedings of the European Conference on Web Services, ECOWS 2004, Erfurt, Germany.

15. Lopes, A., Botelho, L.M. (2005). SEA: a Semantic Web Services Context-aware Execution Agent. AAAI Fall Symposium on Agents and the Semantic Web. Arlington, VA, USA.

16. Martin, D., Burstein, M., Lassila, O., Paolucci, M., Payne, T., McIlraith. S. (2004). Describing Web Services using OWL-S and WSDL. DARPA Markup Language Program.

17. Mitkas, P., Dogac, A. (2002). An Agent Framework for Dynamic Agent Retraining: Agent Academy. eBusiness and eWork 2002, 12th annual conference and exhibition.

18. Nwana, H., Ndumu, D., Lee, L., and Collis, J. (1999). ZEUS: A Tool-Kit for Building Distributed Multi-Agent Systems. Applied Artifical Intelligence Journal, vol. 13, no. 1 pp 129-186.

19. OWL Services Coalition. (2003). OWL-S: Semantic Markup for Web Services. DARPA Markup Language Program.

20. Paolucci, M. and Srinivasan, N. (2004). OWL-S Virtual Machine Project Page. http://projects.semwebcentral.org/projects/owl-s-vm/.

21. Paolucci, M., Soudry, J., Srinivasan, N., Sycara, K. (2004). A Broker for OWL-S Web Services. First International Semantic Web Services Symposium, AAAI Spring Symposium Series.

22. Prud'hommeaux, E. and Seaborne, A. (2004). SPARQL Query Language for RDF. W3C Working Draft. http://www.w3.org/TR/2004/WD-rdf-sparql-query-20041012/

23. Roman, D., Keller, U., Lausen, H. (2004). Web Service Modeling Ontology (WSMO) – version 1.2. Available at http://www.wsmo.org/TR/d2/v1.2/.

24. Sirin, E. (2004). OWL-S API project website. http://www.mindswap.org/2004/owl-s/api/.

An Effective Strategy for the Flexible Provisioning of Service Workflows

Sebastian Stein, Terry R. Payne, and Nicholas R. Jennings

School of Electronics and Computer Science,
University of Southampton,
Southampton, SO17 1BJ, UK
{ss04r,trp,nrj}@ecs.soton.ac.uk

Abstract. Recent advances in service-oriented frameworks and semantic Web technologies have enabled software agents to discover and invoke resources over large distributed systems, in order to meet their high-level objectives. However, most work has failed to acknowledge that such systems are complex and dynamic multi-agent systems, where service providers act autonomously and follow their own decision-making procedures. Hence, the behaviour of these providers is inherently uncertain — services may fail or take uncertain amounts of time to complete. In this work, we address this uncertainty and take an agent-oriented approach to the problem of provisioning service providers for the constituent tasks of abstract workflows. Specifically, we describe an algorithm that uses redundancy to deal with unreliable providers, and we demonstrate that it achieves an 8-14% improvement in average utility over previous work, while performing up to 6 times as well as approaches that do not consider service uncertainty. We also show that our algorithm performs well in the presence of inaccurate service performance information.

1 Introduction

Fuelled by technological advances and the pervasiveness of communication networks, such as the Internet, modern computing devices are increasingly part of complex distributed systems, allowing unprecedented access to a wide range of resources, information sources and remote computing facilities. In this context, service-oriented computing is emerging as a popular approach for allowing autonomous agents to discover and invoke such distributed resources, encapsulated as computer services [1]. In most application scenarios, from computational Grids to automated business processes, multiple services are often combined as part of workflows, thus enabling the consumer to achieve complex goals [2].

Now, a key feature of large distributed systems is that participants are often autonomous agents that follow their own decision-making procedures [3]. This means that the behaviour of service providers is beyond the control of the consumer, and is thus inherently uncertain. This might be manifested by uncertain service durations (e.g., when a provider prioritises service requests from some consumers over others, or when it allocates a variable fraction of its available

J. Huang et al. (Eds.): SOCASE 2007, LNCS 4504, pp. 16–30, 2007.
© Springer-Verlag Berlin Heidelberg 2007

resources to each request), or by unpredictable failures that might occur when a provider is unable (or unwilling) to honour a service request. Such uncertainty is a critical issue for service consumers that need to execute large workflows of interdependent tasks in distributed systems, especially when there are time constraints and when service providers demand remuneration.

So far, this issue has received relatively little attention in the current literature. Instead, most work is concerned with matchmaking techniques that simply identify any single service provider able to fulfil a given goal, based on service descriptions that are assumed to be accurate and truthful [4]. While some workflow languages contain static exception handling methods for dealing with service failures [5], these are inflexible, as they: deal only reactively with problems; need to be specified manually; and usually rely on cooperative services that signal failures and can be rolled back.

To address failures more proactively, some approaches have used quality-of-service measures to place constraints on services or to optimise a weighted sum of various parameters [6], but these require appropriate constraints and weights to be specified by a user. Other research uses utility-theory to choose optimal services in the presence of uncertainty [7]. However, these approaches generally provision only single providers for each task of a workflow, which leads to brittle workflows that are highly vulnerable to single failures. An exception to this is work on highly unreliable services in public-resource computing and peer-to-peer systems, where services are invoked redundantly to increase the overall reliability [8]. While dealing with uncertainty to some extent, such work assumes services to be provided for free, and uses a fixed level of redundancy regardless of the actual reliability of services, time constraints or the value of the workflow.

To deal with this, we previously described a strategy that uses stochastic quality-of-service information about providers to flexibly provision multiple providers for particularly failure-prone tasks in a workflow [9]. In that work, we showed that this redundancy allows the consumer to deal proactively with service uncertainty, and we demonstrated that the strategy worked well even if all available providers were highly unreliable. However, the strategy used a simple estimate of the overall workflow completion time based only on the mean time of each task. Furthermore, we assumed accurate performance information to be available about every task, which may be unrealistic in open and dynamic systems, where such information is often noisy and inaccurate.

In this paper, we substantially extend our previous work in the the following three ways. First, we describe how to calculate and use task duration variance in order to obtain a better estimate for the overall workflow completion time. Second, we investigate the sensitivity of our strategy in the presence of inaccurate service performance information. Third, we show how our strategy performs in more complex environments than previously considered.

The remainder of this paper is structured as follows. In Section 2, we describe the model of a service-oriented system. Then, in Section 3, we outline our extended provisioning strategy. In Section 4, we empirically evaluate the strategy and compare it to our previous work. We conclude in Section 5.

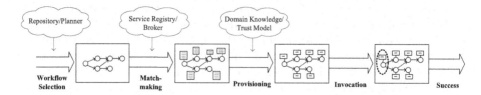

Fig. 1. Lifecycle of a workflow

2   Service-Oriented System Model

In order to provide a formal basis for our work, we briefly describe our system model in this section and relate it to the lifecycle of a workflow, as shown in Fig. 1. As is common in related work, we assume that a service consumer has selected an abstract workflow, describing the types of tasks and their ordering constraints required to achieve some high-level goal (the first stage in Fig. 1). This may originate from a workflow repository or a planner that operates on service types. More formally, we represent such an abstract workflow as a directed acyclic graph $W = (T, E)$, where $T = \{t_1, t_2, \ldots, t_{|T|}\}$ is the set of nodes (the tasks) and $E \in T \leftrightarrow T$ is the set of edges (the ordering constraints). Furthermore, we use a reward function $u \in \mathbb{Z}_0^+ \to \mathbb{R}$ that denotes the utility of completing the workflow at a given time step t. This is based on a maximum reward $u_{\max}$ that is awarded if the workflow is completed no later than a deadline d. If the workflow is late, a cumulative penalty δ is deducted at each time step, until the consumer no longer receives a reward (it does not receive a negative reward — rather, we assume that the workflow has simply failed). Formally, we define u as:

$$u(t) = \begin{cases} u_{\max} & \text{if } t \leq t_{\max} \\ u_{\max} - \delta(t - d) & \text{if } t > d \text{ and } t < d + u_{\max}/\delta \\ 0 & \text{if } t \geq d + u_{\max}/\delta. \end{cases} \tag{1}$$

Once a workflow has been selected, a consumer discovers appropriate service providers for each of the tasks by submitting the abstract task descriptions to a matchmaking process [10] (as shown in the matchmaking stage of Fig. 1). This might be achieved by contacting a service registry or communicating with a broker. We represent this process using a matching function $m \in T \to \wp(S)$, where $S = \{s_1, s_2, \ldots, s_{|S|}\}$ is the set of all service providers. This maps each task t_i to a subset of S to represent the providers that are able to satisfy t_i.

In the next stage, the agent provisions service providers for each task. Here, it makes a decision about how to allocate individual service providers to the constituent tasks of the workflow. We focus on this particular stage, because it allows the agent to use appropriate domain knowledge or information provided by a trust model [11] to make predictions about the feasibility of a workflow, and, where necessary, invest additional resources to provision providers redundantly for particularly failure-prone tasks. In this context, we assume that some limited performance information about the population of providers for each task t_i is known to the consumer[1]:

- $S_i \subseteq S$ is the set of suitable service providers (as given by $m(t_i)$).
- $f_i \in [0,1]$ is the probability that a randomly chosen provider from S_i will default or otherwise fail to deliver a satisfactory result.
- $D_i(t) \in [0,1]$ is the cumulative probability distribution, representing the execution duration of a single service provider (randomly chosen from S_i and assuming it is successful). In particular, $D_i(t)$ is the probability that the provider will take t time steps or less to complete the service.
- $c_i \in \mathbb{R}$ is the cost of invoking a provider from S_i. This may be a financial remuneration or a communication cost.

After provisioning, the consumer starts invoking providers for the tasks of the workflow according to the ordering constraints given by E (the fourth stage in Fig. 1). Here, we assume that providers are only invoked at integer time steps, and, as is common in the Web services domain, this is done *on demand* when providers are required. Hence, the cost for each is paid only at the time of invocation (but regardless of the eventual outcome). When invoked, a provider successfully completes the assigned task t_i with probability $1 - f_i$. The duration of a successful service execution is distributed according to D_i, after which the service consumer is notified of success. When a provider fails, we assume that it does so silently (i.e., no response is given to the consumer). This is realistic in distributed systems, where service providers generally do not reveal their internal state, and where network or machine failures can cause communication losses.

Furthermore, as there may be several matching providers for a task, the consumer can invoke more than one service for this task at the same time. In this case, the consumer has to pay each provider separately, and the task is completed when the earliest service has executed successfully (if any). When all providers seem to have failed, the consumer may invoke new providers for this task. In this case, the consumer will ignore the previously invoked providers and assign the task to the newly provisioned set of services.

Having outlined our basic system model and workflow lifecycle, we now continue to describe our extended provisioning strategy.

3   Flexible Provisioning

In this section, we present a heuristic provisioning strategy that decides dynamically how to provision providers based on the information that is available about each task of the workflow. This strategy is an extension of the work described in [9], and now includes mechanisms for calculating the duration variances of individual tasks and using them to estimate the overall duration distribution of the workflow. In Section 3.1, we begin by giving a high-level overview of our approach, and then, in Sections 3.2 and 3.3, we detail the calculations at the task and workflow level, respectively.

---

[1] In [12] we consider the case where more detailed information about individual providers is available.

3.1 Provisioning Problem

As discussed in more detail in our previous work [9], we use two forms of redundancy to deal with uncertainty: parallel and serial provisioning. More specifically, for each task t_i, we determine a number of providers (n_i) to invoke in parallel. Provisioning multiple providers in such a way increases the probability of success for that task, and also decreases its expected duration (as the consumer can proceed as soon as one of the providers is successful). Furthermore, to deal with the case where none of the providers is successful, we determine a maximum waiting time (w_i) before the consumer invokes a new set of n_i service providers. However, using redundancy in such a way increases the cost for the overall workflow, and, hence, we aim to maximise the expected net profit (denoted $\bar{u}(\boldsymbol{n}, \boldsymbol{w})$):

$$\max_{\boldsymbol{n},\boldsymbol{w} \in \mathbb{N}^{|T|}} \bar{u}(\boldsymbol{n}, \boldsymbol{w}), \qquad (2)$$

where $\boldsymbol{n}$ and $\boldsymbol{w}$ are vectors, whose ith elements correspond to the number of parallel providers (n_i) and the waiting time (w_i) of each task t_i of the workflow. Furthermore, we define the expected net profit $\bar{u}(\boldsymbol{n}, \boldsymbol{w})$ as the difference of the expected reward $\bar{r}(\boldsymbol{n}, \boldsymbol{w})$ and the expected cost $\bar{c}(\boldsymbol{n}, \boldsymbol{w})$ when provisioning the workflow using the vectors $\boldsymbol{n}$ and $\boldsymbol{w}$:

$$\bar{u}(\boldsymbol{n}, \boldsymbol{w}) = \bar{r}(\boldsymbol{n}, \boldsymbol{w}) - \bar{c}(\boldsymbol{n}, \boldsymbol{w}). \qquad (3)$$

However, even the calculation of this objective function for any given vectors $\boldsymbol{n}$ and $\boldsymbol{w}$ is intractable, because it requires the calculation of the overall duration distribution (this is known to be a $\#P$-complete problem [13]). For this reason, we use a heuristic strategy that estimates the expected reward and cost of a random initial choice for $\boldsymbol{n}$ and $\boldsymbol{w}$, and then performs steepest-ascent hill-climbing to find a good solution. More specifically, at each iteration, our hill-climbing algorithm generates a large set of neighbours of the current best choice for $\boldsymbol{n}$ and $\boldsymbol{w}$ by performing small changes to it. To generate each neighbour, the algorithm picks one component of either vector (n_i or w_i), then increases or decreases it by either 1 or a random amount. This is done systematically for each task, thus producing $8 \cdot |T|$ neighbours in total. Each neighbour is then evaluated, the best is chosen and this process is repeated until no more improvements can be made.

Now, in our previous work, we used a simple calculation for estimating the expected reward $\bar{r}(\boldsymbol{n}, \boldsymbol{w})$. In particular, our mechanism assumed task durations to be deterministic and then used the longest path in the workflow to calculate an overall, deterministic duration $\tilde{t}$ (conditional on success). Hence, it estimated the expected reward as $\tilde{r} = p \cdot u(\tilde{t})$, where p is the overall success probability of the workflow, and $u(\tilde{t})$ is the reward at time $\tilde{t}$ (omitting the parameters $\boldsymbol{n}$ and $\boldsymbol{w}$ for brevity). A major shortcoming of this approach is that it does not consider the variance of task durations at all, and so often overestimates the expected reward. To address this, we artificially increased our estimate for $\tilde{t}$ by a constant 20%, but such an approach still does not consider variance, and is not guaranteed to work well in all environments.

To overcome this, we now use an improved heuristic that does not assume a deterministic duration, but rather estimates the probability distribution of the workflow duration. To this end, we estimate the expected profit as:

$$\tilde{u} = p \int_0^\infty d_W(x)u(x)\,\mathrm{d}x \;-\; \tilde{c}, \tag{4}$$

where $d_W(x)$ is a probability density function that estimates the overall duration of the workflow, and $\tilde{c}$ is an estimate of the expected overall cost (using a continuous probability function allows us to derive a simple and concise solution in closed form). This equation is central to our heuristic strategy, and, in the following two sections, we describe how to calculate its components.

3.2   Local Task Calculations

In order to solve Equation 4, we first calculate four parameters for each task t_i in the workflow: its success probability, expected cost, expected duration and duration variance. These are discussed in more detail below.

Success Probability (p_i): This is the probability that the task will be successful, regardless of the eventual finishing time. Because we assume that the consumer will continue invoking providers until the task is completed (with the given waiting time w_i between invocations), it is the probability that at least one provider from the set S_i is successful within its allocated time:

$$p_i = 1 - (1 - (1 - f_i) \cdot D_i(w_i))^{|S_i|}. \tag{5}$$

Expected Cost ($\bar{c}_i$): This is expected overall cost that the consumer will spend on the task. To calculate it, we let $\hat{f}_i = (1-(1-f_i)\cdot D_i(w_i))^{n_i}$ be the probability that a single invocation of n_i parallel providers is unsuccessful within their given waiting time, $m = \lfloor |S_i|/n_i \rfloor$ the maximum number of such invocations (with n_i providers), and $r = |S_i| \bmod n_i$ the number of service providers available for the last invocation. The expected cost is then:

$$\bar{c}_i = n_i c_i \cdot \frac{1 - \hat{f}_i^m}{1 - \hat{f}_i} + \hat{f}_i^m c_i r. \tag{6}$$

Expected Duration ($\bar{t}_i$): This is the expected time the providers will take to complete the task (conditional on overall success). Here, we let $\mu_i = \hat{D}_i(w_i)^{-1}$ $\sum_{k=1}^{w_i} k \cdot (\hat{D}_i(k) - \hat{D}_i(k-1))$ be the mean time to success of a single successful invocation of n_i service providers (where $\hat{D}_i(x) = 1 - (1 - (1 - f_i) \cdot D_i(x))^{n_i}$ is the cumulative non-conditional probability that at least one out of n_i services has finished successfully by time x), λ_i the corresponding mean time for the final invocation of r providers, and $\check{f}_r$ the failure probability of that invocation (calculated analogously to $\hat{f}_i$). Hence, we calculate the expected duration as:

$$\bar{t}_i = \frac{1}{p_i}(\mu_i(1 - \hat{f}_i^m)) + w_i \frac{\hat{f}_i - m\hat{f}_i^m + (m-1)\hat{f}_i^{m+1}}{1 - \hat{f}_i} + \hat{f}_i^m(1 - \check{f}_r)(\lambda_i + m w_i)). \tag{7}$$

Variance of Duration (σ_i^2): This is the variance of the task duration. To calculate it, we let C_i be a random variable representing the duration of the task, conditional on its success. Then, $E(C_i)$ is its expected value (note that $E(C_i) = \bar{t}_i$), $E(C_i^2)$ is its expected square, and $VAR(C_i)$ is its variance. Hence,

$$\sigma_i^2 = VAR(C_i) = E(C_i^2) - E(C_i)^2 = E(C_i^2) - \bar{t}_i^2. \tag{8}$$

In order to obtain $E(C_i^2)$, we consider two cases: (1) the task is completed successfully within the first m invocations, and (2) it is completed during the last invocation of r providers (if $r \neq 0$). We denote the durations in each case by the random variables A_i and B_i, respectively. This allows us to treat both cases separately, and, letting P_A and P_B be the respective probabilities of each case occurring (conditional on overall success), we write $E(C_i^2)$ as:

$$E(C_i^2) = P_A E(A_i^2) + P_B E(B_i^2) = \frac{1 - \hat{f}_i^m}{1 - \check{f}_r \hat{f}_i^m} E(A_i^2) + \frac{\hat{f}_i^m (1 - \check{f}_r)}{1 - \check{f}_r \hat{f}_i^m} E(B_i^2). \tag{9}$$

Next, we note that each duration is divided into a time period spent waiting during any unsuccessful invocations (we denote these as A_{Wi} and B_{Wi}), and a period spent during the final invocation until the first provider is successful (we denote these as A_{Di} and B_{Di}). We note that these are independent in our model, and treat case (1) first:

$$
\begin{aligned}
E(A_i^2) &= VAR(A_i) + E(A_i)^2 \\
&= VAR(A_{Wi}) + VAR(A_{Di}) + (E(A_{Wi}) + E(A_{Di}))^2 \\
&= E(A_{Wi}^2) + E(A_{Di}^2) + 2E(A_{Wi})E(A_{Di}).
\end{aligned}
\tag{10}
$$

Now, $E(A_{Di}) = \mu_i$, and $E(A_{Di}^2)$ can be similarly calculated by multiplying the term inside the summation by k^2 instead of k. Furthermore, we obtain $E(A_{Wi})$ from Equation 7, and then derive $E(A_{Wi}^2)$ in a similar way:

$$E(A_{Wi}) = \frac{w_i(\hat{f}_i - m\hat{f}_i^m + (m-1)\hat{f}_i^{m+1})}{(1 - \hat{f}_i)(1 - \hat{f}_i^m)} \tag{11}$$

$$E(A_{Wi}^2) = \frac{(1 - \hat{f}_i)w_i^2}{1 - \hat{f}_i^m} \sum_{k=0}^{m-1} k^2 \hat{f}_i^k \tag{12}$$

$$
\begin{aligned}
&= w_i^2(\hat{f}_i + \hat{f}_i^2 - m^2\hat{f}_i^m - (2m + 1 - 2m^2)\hat{f}_i^{m+1} \\
&\quad + (2m - 1 - m^2)\hat{f}_i^{m+2})(1 - \hat{f}_i^m)^{-1}(1 - \hat{f}_i)^{-2}.
\end{aligned}
$$

Treating case (2) next, we calculate $E(B_i^2)$ analogously to Equation 10. This is easier, as the waiting time B_{Wi} is constant (mw_i). Hence,

$$
\begin{aligned}
E(B_i^2) &= E(B_{Wi}^2) + E(B_{Di}^2) + 2E(B_{Wi})E(B_{Di}) \\
&= (mw_i)^2 + E(B_{Di}^2) + 2mw_i E(B_{Di}).
\end{aligned}
\tag{13}
$$

The remaining terms, $E(B_{Di})$ and $E(B_{Di}^2)$, are calculated as $E(A_{Di})$ and $E(A_{Di}^2)$, discussed above. Combining all terms to solve Equation 8 gives us a closed form for calculating the variance of the duration of a given task. With these parameters for each task, we now continue to describe how they are combined over the entire workflow to solve our heuristic function (Equation 4).

3.3   Workflow Calculations

In order to solve Equation 4, we require the overall success probability p, a distribution for the workflow duration $d_W(x)$, and the estimated overall cost $\tilde{c}$. The success probability p is simply the probability that all tasks are eventually successful, and the estimated total cost $\tilde{c}$ is the sum of all task costs, each multiplied by the probability that the task is ever reached (denoted r_i):

$$p = \prod_{\{i|t_i \in T\}} p_i. \tag{14}$$

$$\tilde{c} = \sum_{\{i|t_i \in T\}} r_i \bar{c}_i \tag{15}$$

$$r_i = \begin{cases} 1 & \text{if } \forall t_j \cdot ((t_j \mapsto t_i) \notin E) \\ \prod_{\{j|(t_j \mapsto t_i) \in E\}} p_j & \text{otherwise.} \end{cases} \tag{16}$$

To estimate the workflow duration function $d_W(x)$, we use a technique from operations research [14]. In particular, we consider the *critical path* of the workflow (i.e., the path that maximises the sum of all mean task durations along it) and obtain the sum of all mean task durations (λ_W) and variances on it (v_W). Exploiting the central limit theorem, we then approximate the duration of the workflow using a normal distribution with mean λ_W and variance v_W:

$$d_W(x) = \frac{1}{\sqrt{v_W 2\pi}} e^{-\frac{(x-\lambda_W)^2}{2v_W}}. \tag{17}$$

Now, to solve Equation 4, we let $D_W(x) = \int_{-\infty}^{x} d_W(y)\, dy$ be the cumulative probability function[2] of $d_W(x)$, we let $D_{\max} = D_W(t_{\max})$ be the probability that the workflow will finish within the deadline $t_{\max}$ and $D_{\text{late}} = D_W(t_0) - D_W(t_{\max})$ the probability that the workflow will finish after the deadline but no later than time $t_0 = u_{\max}/\delta + t_{\max}$ (both conditional on overall success).

Next, we consider three distinct cases, based on the form of $u(t)$ (Equation 1). First, the workflow may finish within the deadline $t_{\max}$ — this happens with probability $D_{\max}$ and results in the full reward, $u_{\max}$. Second, the workflow may finish after t_0 — this happens with probability $1 - D_W(t_0)$, and here the consumer receives no reward (and so we can ignore it). Finally, the workflow may finish between these two times, which happens with probability D_{late}. Because $u(t)$ is linear on this interval, we calculate the expected reward in this case by

---

[2] This common function is usually approximated numerically. In our work, we use the SSJ library (http://www.iro.umontreal.ca/~simardr/ssj).

applying $u(t)$ to the mean time on the interval, which we denote by $\bar{t}_{late}$. Hence, we re-write Equation 4, concluding our heuristic utililty function:

$$\tilde{u} = p \cdot (D_{max} \cdot u_{max} + D_{late} \cdot u(\bar{t}_{late})) - \tilde{c} \tag{18}$$

$$\text{with} \quad \bar{t}_{late} = \frac{1}{D_{late}} \int_{t_{max}}^{t_0} d_W(x) x \, dx$$

$$= \lambda_W + (e^{\frac{-(t_{max} - \lambda_W)^2}{2 v_W}} - e^{\frac{-(t_0 - \lambda_W)^2}{2 v_W}}) \cdot \frac{\sqrt{v_W}}{D_{late} \cdot \sqrt{2\pi}}. \tag{19}$$

Using this heuristic function, it is now possible to use steepest-ascent hill-climbing as described at the beginning of this section. Through observations, we have seen that our hill-climbing algorithm quickly converges to a good solution[3]. In particular, the heuristic function $\tilde{u}$ can be solved efficiently in quadratic time. The bottleneck here is the calculation for Equations 15 and 16. However, after the initial calculation, only small adjustments need to be made at each iteration of the hill-climbing procedure, further reducing the run-time of calculating $\tilde{u}$. In this case, it is bounded by the critical path problem required to obtain λ_W and v_W used in Equation 19, which has a run-time in $O(|T| + |\mathcal{E}|)$ where $|T|$ is the number of tasks in the workflow and $|\mathcal{E}|$ the number of non-transitive edges[4].

Having concluded our discussion of the improved provisioning strategy, we now describe a set of empirical experiments to evaluate our work.

4   Empirical Evaluation

To investigate the performance of our strategy and compare it to our previous work, we conduct a set of empirical experiments using the same methodology as described in [9]. To summarise briefly, we test our strategy by randomly generating a set of service providers and a single workflow according to a set of controlled variables that specify a particular environment. Then, we provision the workflow using our strategy and simulate its execution, recording the overall net profit achieved. This is repeated 1000 times for each environment to achieve statistical significance (we give all results with 95% confidence intervals and carry out ANOVA and two-sample t-tests as appropriate, both at the 99% confidence level). Throughout all experiments, we vary the failure probability of providers (as given by f_i) to evaluate our strategy in environments with varying levels of service unreliability, keeping all other variables constant for consistency. To provide a benchmark, we compare our results to a *naïve* strategy that provisions a single provider for each task of the workflow. This strategy represents currently prevalent approaches that ignore any service uncertainty and thus rely only on functional service descriptions to find any suitable provider for each task.

---

[3] On average, around six iterations are needed per task in the workflow. During the empirical evaluation of our algorithm (see Section 4), a solution was typically found within 250ms (10 tasks) or 5s (50 tasks) on a 3GHz Pentium 4 with 1GB RAM.

[4] We also assume that the probability density functions of service durations and expected values (e.g., μ_i in Equation 7) can be efficiently calculated or approximated.

In the remainder of this section, we first test our improved strategy using the same experimental setup as in our previous work to allow a direct comparison (Section 4.1). Because that setup considers relatively small workflows with homogeneous tasks, we then examine a more complex scenario with heterogeneous tasks and larger workflows in Section 4.2. Finally, in Section 4.3, we evaluate the performance of our strategy in the presence of inaccurate information.

4.1 Small Workflows

In order to compare the improved strategy to our previous work, we first consider the same environments as described in [9] (however, for consistency, we repeat all experiments). Here, workflows consist of 10 tasks in a strict sequence, they have a deadline of $d = 400$ time steps, a maximum utility $u_{\max} = 1000$ and a penalty of $\delta = 10$ per time step. Furthermore, there are 1000 providers able to satisfy each task. Each provider has a cost $c_i = 10$ and its duration distribution is a gamma distribution with parameters $k = 2$ and $\theta = 10$ (hence, the duration distribution for each provider is $D_i(t) = \gamma(k, t/\theta)\Gamma(k)^{-1}$).

Fig. 2(a) shows the results of our improved strategy (*improved flexible*), our previous strategy (*old flexible*) and the *naïve* strategy. Clearly, the performance of all three strategies is significantly different. Averaged over all failure probabilities, the *naïve* strategy achieves a net profit of only 106.00 ± 5.95, and, in fact, begins making an overall loss when the failure probability drops to 0.3 or lower. As discussed in previous work, our *old flexible* strategy clearly improves on this, yielding an average net profit of 474.91 ± 7.77. The results also indicate that our modified *improved flexible* strategy further yields a significant improvement in overall performance with an average net profit of 515.02 ± 4.02. At all failure probabilities from 0.1–0.7, the strategies achieve significantly different amounts of net profit, with the *improved flexible* strategy outperforming all others. At 0.8 and beyond, the difference between the two flexible strategies becomes less pronounced due to the overall low net profit, and it is no longer significant. However, they still outperform the *naïve* strategy. When the failure probability is 0, all strategies perform equally well, as a single provider for each task is always sufficient to complete the workflow in time.

To summarise, averaged over the failure probabilities discussed here, our *improved flexible* strategy achieves a $385.87\% \pm 8.69\%$ improvement over the *naïve* strategy, and an $8.44\% \pm 2.21\%$ improvement over the *old flexible* strategy.

4.2 Large Workflows

To show that our strategy performs well in more complex environments, we perform another set of experiments with workflows consisting of 50 tasks and heterogeneous task parameters. More specifically, for each experimental run, we first generate a workflow by randomly populating an adjacency matrix until 25% of all possible edges have been added (ignoring any that would result in a cyclic graph). This ensures that the workflow contains a considerable number of parallel tasks. We also assume there are 100 providers for each task, the deadline is $d = 1000$, the maximum reward $u_{\max} = 1000$ and the penalty $\delta = 1$.

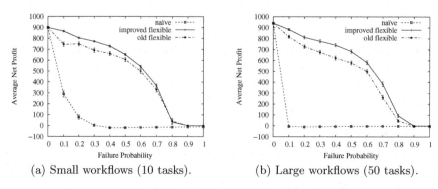

(a) Small workflows (10 tasks).   (b) Large workflows (50 tasks).

Fig. 2. Net profit of flexible strategies

In order to vary the performance characteristics of the tasks in the workflow, we randomly assign each to one of seven different types of tasks with varying costs (c_i) and duration distributions (D_i), as shown in Table 1. For every experimental run, we also attach a failure probability to each type (to determine f_i) that is drawn from a beta distribution with parameters $\alpha = 10 \cdot f$ and $\beta = 10 - \alpha$, where f is the average failure probability of the environment (unless $f = 0$ or $f = 1$, in which case all tasks have the same failure probability). This adds further variance to the tasks, while ensuring that the overall average failure probability over all runs is close to f.

Fig. 2(b) shows similar results to those described in the previous section. When providers are always reliable (the failure probability is 0), there is no significant difference between the strategies. However, as soon as providers begin to fail, the *naïve* strategy begins to perform poorly and make an overall loss. The two flexible strategies achieve far better results and avoid making a loss in any environment. As before, our *improved flexible* strategy outperforms the *old flexible* strategy in most environments, except when the overall failure probability reaches 0.9, when there is no more significant difference between them.

When averaging over all failure probabilities, the *naïve* strategy achieves an average net profit of 81.53 ± 5.12, the *old flexible* achieves 470.21 ± 7.44 and the *improved flexible* achieves 536.12 ± 7.46. This means that our *improved flexible*

Table 1. Service types used to test complex workflows

Type	Cost	Duration	Mean	Variance
T_1	0.1	Gamma(1,0.1)	0.1	0.01
T_2	0.1	Gamma(1,10)	10	100
T_3	1	Gamma(5,1)	5	5
T_4	1	Gamma(5,10)	50	500
T_5	2	Gamma(10,1)	10	10
T_6	2	Gamma(10,5)	50	250
T_7	2	Gamma(100,0.1)	10	1

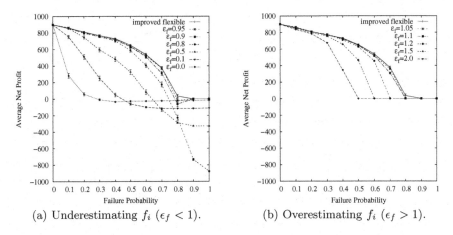

(a) Underestimating f_i ($\epsilon_f < 1$).   (b) Overestimating f_i ($\epsilon_f > 1$).

Fig. 3. Effect of incorrect failure probabilities

strategy achieves a $557.60\% \pm 11.10\%$ improvement over the *naïve* strategy and a $14.02\% \pm 2.24\%$ improvement over the *old flexible* strategy.

4.3   Sensitivity Analysis

In order to evaluate the performance of our strategy in the presence of inaccurate information, we follow the same experimental setup as in Section 4.1, but now systematically introduce errors into the information that is available to a service consumer following the *improved flexible* strategy. To this end, we first evaluate the effect of relying on inaccurate failure probabilities, and then examine the impact of inaccurate service duration information. In both cases, we expect the performance of our strategy to decrease as the information becomes less accurate. However, because we rely on heuristic estimates, we anticipate that small inaccuracies will have little overall impact on the performance of the strategy.

In our first set of experiments, we consider the case where the consumer *underestimates* the failure probability of service providers. Hence, we multiply the actual values for the failure probabilities f_i by a scalar $\epsilon_f < 1$ to provide an inaccurate input to the *improved flexible* strategy. The results for various values of ϵ_f are shown in Fig. 3(a). In most cases, the average net profit gained by the strategy degrades gracefully as the performance information becomes more inaccurate. In fact, when the (true) failure probability is low in the environment (up to around 0.3), the strategy does well even if the information is up to 90% inaccurate (i.e., $\epsilon_f = 0.1$). However, when the failure probability rises to 0.7 and beyond, the impact of inaccurate information becomes more detrimental to the performance of the strategy. This is particularly evident when $\epsilon_f = 0.8$, which results in a large net loss at high failure probabilities. This is because the strategy provisions a large number of providers in parallel without detecting that the workflow is infeasible (and thus, it loses its high investment). Perhaps surprisingly, when information becomes even more inaccurate at high failure

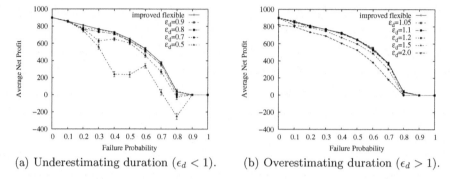

(a) Underestimating duration ($\epsilon_d < 1$).   (b) Overestimating duration ($\epsilon_d > 1$).

Fig. 4. Effect of incorrect failure probabilities

probabilities, the consumer begins to make smaller losses again. This is due to the strategy provisioning less providers in parallel and therefore losing less of its investment when the workflow eventually fails. Despite the special case when $\epsilon_f = 0.8$, the results are promising and show that small inaccuracies in the information (up to 10%) have little or no effect on our strategy. In most other cases, performance degrades gracefully as the information becomes less accurate.

Next, we are interested in the trends resulting from *overestimating* the failure probability of service providers. Hence, we now multiply the failure probabilities by a scalar $\epsilon_f > 1$ to provide an inaccurate input to our strategy (using a failure probability of 1 whenever $f_i \cdot \epsilon_f > 1$). The results of this are shown in Fig. 3(b). Not surprisingly, the performance of the strategy simply degrades as the perceived failure probability rises. Because its behaviour is more conservative when it overestimates the failure probability of providers (it will provision unnecessarily many providers), it never makes a long-term loss. These results show that our strategy performs well, even when it significantly overestimates failure probabilities. In fact, the overall performance degrades only slightly when the failure probability is overestimated by 10% ($\epsilon_f = 1.1$). Even at 20% ($\epsilon_f = 1.2$), the performance is extremely good, and at 50% the strategy still performs reasonably well compared to the case with accurate information.

Apart from the failure probabilities, our strategy also relies on probability density functions for the duration of a service execution. Because these will most likely be based on past observations and can be subject to noise, we now examine the effect of inaccurate information about these functions. Here, we multiply the scale parameter θ of the underlying gamma distribution by a scalar ϵ_d to yield an inaccurate duration distribution. By varying the scale parameter, we ensure that the mean of the distribution is varied proportionally with ϵ_d (e.g., when $\epsilon_d = 0.5$, the consumer estimates the mean service execution time to be half of the true value), while the overall shape of the distribution stays the same.

Again, we first consider the case of *underestimating* the duration of service providers ($\epsilon_d < 1$). The results are shown in Fig. 4(a). Here, the strategy handles an error of up to 20% ($\epsilon_d = 0.8$) very well with only a marginal performance decrease. Even when the error rises to 30% ($\epsilon_d = 0.7$), the performance comes close to the accurate information case. However, as the information becomes even more inaccurate, the strategy performs increasingly badly. Also, the strategy behaves more erratically at the same time — occasionally, the average net profit at a given level of inaccuracy increases as the failure probability rises (this is because the strategy constantly varies the balance between parallel and serial invocations, the latter of which is more susceptible to wrong duration estimates).

Finally, Fig. 4(b) shows the corresponding results when the consumer *overestimates* the service duration. Here, the performance again degrades slowly as the error rises. This is because the agent allocates unnecessarily long waiting times to the providers or provisions parallel providers when this is not needed. However, the loss in performance is clearly very small. This is because the consumer will occasionally wait longer than required or incur extra expenditure by provisioning parallel providers, but in many cases, the providers will simply complete their services earlier than anticipated and the consumer will be able to continue the workflow immediately and without penalty.

To conclude the sensitivity analysis, the results presented in this section show that our strategy is robust to small and moderate inaccuracies. In all cases, it performs well when the information provided is within 10% of the true value, and often errors up to 20% and 30% lead to only marginal decreases in performance, especially when the consumer is overly pessimistic (i.e., when it overestimates the failure probability or duration of services). Overall, performance generally degrades gracefully as larger errors are introduced into the information that is known about providers (until they are too large to be of any value to the consumer — e.g., as ϵ_d reaches 0.5).

We also identified one case where underestimating the failure probability of providers can lead to poor performance. However, this only occurs in very specific scenarios when providers are highly unreliable and when the error in information is a significant 20%. Hence, our strategy may benefit from identifying these conditions in advance, and we will consider this in future work. Nevertheless, the overall results presented here are promising, showing that our strategy is applicable even in environments where completely accurate performance information is unavailable (as will be typical in any large dynamic multi-agent system).

5   Conclusions

In this paper, we have extended our previous work by taking into consideration the variance of service duration times. This is important to consider when predicting the overall completion time of a workflow, and in empirical experiments, we have shown that our new strategy performs significantly better than our previous work. We have also given more extensive results in a variety of settings and shown that our strategy copes well in environments where performance information is not accurate.

Considering these results, our work is highly applicable in realistic scenarios, where software agents autonomously execute large workflows in service-oriented systems. Examples of such scenarios include: the Grid domain, where expensive and time-intensive data processing services are often required as part of complex workflows; peer-to-peer systems, where a vast amount of cheap, but usually unreliable providers offer their services; and e-commerce applications, where the consumers often face highly valued workflows with strict deadlines (e.g., extensive supply-chains or large commercial projects).

In future work, we will extend our strategy to handle more advanced negotiation mechanisms rather than the fixed price model we use at the moment. Such mechanisms have been widely used in the context of multi-agent systems, and are now increasingly being employed in service-oriented systems to allow consumers and providers to reach service-level agreements prior to invocation. We also plan to improve the adaptivity of our strategy, utilising information about workflow progress to dynamically alter the provisioning of services.

Acknowledgements. This work was funded by the Engineering and Physical Sciences Research Council (EPSRC) and a BAE Systems studentship.

References

1. Huhns, M.N., Singh, M.P.: Service-oriented computing: Key concepts and principles. IEEE Internet Comput. **9**(1) (2005) 75–81
2. Mandell, D.J., McIlraith, S.A.: Adapting BPEL4WS for the semantic web: The bottom-up approach to web service interoperation. In: Proc. 2nd Int. Semantic Web Conf. (ISWC03), Sanibel Island, USA, Springer (2003) 227–241
3. Jennings, N.R.: An agent-based approach for building complex software systems. Comm. ACM **44**(4) (2001) 35–41
4. McIlraith, S.A., Son, T.C.: Adapting Golog for Composition of Semantic Web Services. In: Proc. 8th Int. Conf. on Knowledge Representation and Reasoning (KR2002), Toulouse, France, Morgan Kaufmann (2002) 482–493
5. Curbera, F., Khalaf, R., Mukhi, N., Tai, S., Weerawarana, S.: The next step in Web services. Comm. ACM **46**(10) (2003) 29–34
6. Zeng, L., Benatallah, B., Dumas, M., Kalagnanam, J., Sheng, Q.Z.: Quality driven web services composition. In: Proc. 12th Int. World Wide Web Conf. (WWW'03), Budapest, Hungary, ACM Press (2003) 411–421
7. Collins, J., Bilot, C., Gini, M., Mobasher, B.: Decision processes in agent-based automated contracting. IEEE Internet Computing **5**(2) (2001) 61–72
8. Anderson, D.P., Cobb, J., Korpela, E., Lebofsky, M., Werthimer, D.: SETI@home: an experiment in public-resource computing. Comm. ACM **45**(11) (2002) 56–61
9. Stein, S., Jennings, N.R., Payne, T.R.: Flexible provisioning of service workflows. In: Proc. 17th Eur. Conf. on AI (ECAI-06), Riva, Italy, IOS Press (2006) 295–299
10. Decker, K., Williamson, M., Sycara, K.: Matchmaking and brokering. In: Proc. 2nd Int. Conf. on Multi-Agent Systems (ICMAS'96), Kyoto, Japan. (1996) 432–433
11. Teacy, W.T.L., Patel, J., Jennings, N.R., Luck, M.: TRAVOS: Trust and reputation in the context of inaccurate information sources. JAAMAS **12**(2) (2006) 183–198
12. Stein, S., Jennings, N.R., Payne, T.R.: Provisioning heterogeneous and unreliable providers for service workflows. In: Proc. 6th Int. Conf. on Autonomous Agents and Multiagent Systems (AAMAS07), Honolulu, Hawai'i. (2007)
13. Hagstrom, J.N.: Computational complexity of PERT problems. Networks **18** (1988) 139–147
14. Malcolm, D.G., Roseboom, J.H., Clark, C.E., Fazar, W.: Application of a technique for research and development program evaluation. Oper. Res. **7**(5) (1959) 646–669

Using Goals for Flexible Service Orchestration
A First Step*

M. Birna van Riemsdijk and Martin Wirsing

Ludwig-Maximilians-Universität München, Germany
{riemsdijk,wirsing}@pst.ifi.lmu.de

Abstract. This paper contributes to a line of research that aims to apply agent-oriented techniques in the field of service-oriented computing. In particular, we propose to use goal-oriented techniques from the field of cognitive agent programming for service orchestration. The advantage of using an explicit representation of goals in programming languages is the flexibility in handling failure that goals provide. Moreover, goals have a close correspondence with declarative descriptions as used in the context of semantic web services. This paper now presents first steps towards the definition of a goal-based orchestration language that makes use of semantic matchmaking. The orchestration language we propose and its semantics are formally defined and analyzed, using operational semantics.

1 Introduction

This paper contributes to a line of research that aims to apply agent-oriented techniques in the field of service-oriented computing. Services are generally defined as autonomous, platform-independent computational entities that can be described, published, and discovered. An important concern in service-oriented computing is how services can be composed in order to solve more complex tasks. One way to go about this, is to use a so-called *orchestration language* such as WS-BPEL [10] or Orc [7], by means of which one can specify an executable pattern of service invocations. Another important issue in the context of services is dealing with *failure* [14,6]. Especially when services are discovered at run-time, one needs to take into account that a service might not do exactly what one had asked for, or that a particular orchestration does not yield the desired result.

We argue that the agent community has something to offer to the services community, as agents are meant to be capable of flexible action in dynamic environments. Being capable of flexible action means that the agent should be able to cope with failure, and should respond adequately to changed circumstances. The idea is now that using agent-oriented techniques in orchestration languages could potentially yield more adaptive and flexible orchestrations.

In this paper, we focus in particular on the usage of *(declarative) goals* as is common in agent programming (see, e.g., [22,5,20]), for flexible service orchestration. Goals as used in agent programming describe situations that the agent

---

* This work has been sponsored by the project SENSORIA, IST-2005-016004.

J. Huang et al. (Eds.): SOCASE 2007, LNCS 4504, pp. 31–48, 2007.

wants to reach. The use of an explicit representation of goals in a programming language provides for added flexibility when it comes to failure handling, as the fact that a goal represents a desired state, can be used to check whether some plan for achieving a goal has failed. This is then combined with a mechanism for specifying which plan may be used for achieving a certain goal in certain circumstances, which also allows for the specification of multiple plans for achieving a certain goal.

When considering the application of goal-oriented techniques to services, especially *semantic web services* seem to have a natural relation with goals. The idea of semantic web services is to endow web services with a declarative description of what the service has to offer, i.e., a declarative description of the semantics of the service. This then allows *semantic matchmaking* [16] between the declarative description of a service being sought, which in our case would correspond with a goal of an agent, and a description of the service being offered. In fact, the WSMO framework for web service discovery refers explicitly to semantic matching between goals and web services [19].

This paper now describes first steps towards the design of a *goal-based orchestration language that makes use of semantic matchmaking*, building on research on goal-oriented agent programming, orchestration, and semantic web services. This orchestration language and its semantics are formally defined and analyzed. The orchestration language and the description of services are based on propositional logic. Being a first step towards combining cognitive agent programming languages and orchestration languages, the relative simplicity of propositional logic allows us to focus on the essential aspects of an agent-based orchestration language.

Proposals for combining agent-oriented and service-oriented approaches are appearing increasingly often in the literature. In the CooWS platform [4], for example, an agent-based approach to procedural learning is used in the context of web services. Other approaches focus on communication protocols [13,2]. The relation between goals as used in agent programming and semantic web services has also been pointed out in [9], in which an architecture is described that focuses on the translation of high-level user goals to lower-level goals that can be related to semantic service descriptions.

To the best of our knowledge, however, this is the first proposal for a formally defined language for goal-based orchestration. Flexibility in handling failure is an issue that is well-recognized in service composition, and it is exactly this flexibility that goal-based techniques can provide to orchestration languages. The main technical contribution of this paper is that we provide a clean and formally defined way of combining service-oriented and agent-oriented models of computation.

2   Syntax

In this section, we describe the syntax and informal semantics of our goal-oriented orchestration language. The way we go about combining goal-oriented techniques and orchestration languages, is that we take (a family of) cognitive agent programming languages as a basis, and incorporate into these constructs for service orchestration. The family of agent programming languages on which we build are variants of the language 3APL as also reported in [20], and we moreover draw inspiration from the language AgentSpeak(L) [18]. The orchestration language that we take as a basis is the language Orc [7].

In (goal-oriented) agent programming languages, agents generally have a belief base, representing what the agent believes to be the case in the world, and a goal base, representing the top-level goals of the agent. Agents execute plans in order to achieve their goals. These plans, broadly speaking, consist of actions that the agent can execute, and so-called subgoals. In order to achieve these subgoals, an agent needs to select plans, just like it selects plans to achieve its top-level goals. Selecting plans for (sub)goals is done by means of so-called *plan selection rules* that tell the agent which plan they may use for achieving which goal, given a certain state of the world.

The general idea of our goal-oriented orchestration language now is that an agent may use not only actions that it can execute itself for achieving its goals, but it can also call services. We thus extend the language of plans with a construct for calling services. These services may be called directly using the service name, or may be discovered, based on the goal that the agent wants to achieve. One of the main technical issues that arises when modifying an agent programming language in this way, is that it has to be determined how to handle the results that are returned by services in a way that is in line with the (goal-oriented) model of computation of agent programming languages. This will be explained in more detail in the sequel.

In order to illustrate our approach, we use a very simple example scenario that is adapted from [23]. In the car repair scenario, the car's diagnostic system reports a failure so that the car is no longer drivable. Depending on the problem, the car may be repaired on the spot, or it might have to be towed to a garage. The car's discovery system then calls a repair service in order to try to repair the car on the spot. Alternatively, it may identify and contact garages and towing truck services in the car's vicinity in order to get the car towed to a garage.

In Section 2.1, we define how services are described, and in Section 2.2 we define the syntax of the orchestration language.

2.1 Service Descriptions

The way we describe services is based on the Ontology Web Language for Services (OWL-S) [12] which seeks to provide the building blocks for encoding rich semantic service descriptions. In particular, we focus on the description of functional properties of services. According to OWL-S, these can be described in terms of *inputs, outputs, preconditions*, and *effects* (so-called IOPEs). The idea is that a service is "a process which requires inputs, and some precondition to be valid, and it results in outputs and some effects to become true" [12].

The inputs description specifies what kind of inputs the service accepts from a caller, and the outputs description specifies what the service may return to the caller. Preconditions are conditions on the state of the world, which need to hold in order for the service to be able to execute successfully. Effects describe what the service can bring about in the world, i.e., the effects description is like a description of post-conditions.

In this paper, the inputs description is a set of atoms. The set of atoms represents what kind of formulas the service is able to handle (similar to a type specification). The outputs description is a set of propositional formulas or the special atom *failure*. Informally, the idea is that the formulas in the outputs description represent alternative possible outputs, where the actual output may also be a combination of these alternatives (see Definition 9 for a precise specification). The effects are described in a similar way.

Below, we formally define service descriptions. A service description has a name, and inputs, outputs, preconditions, and effects. The name of the service may not be the reserved name d. The name d is used to express in plans that a service should be discovered.

Definition 1 *(service description).* Throughout this paper we assume a language of propositional logic $\mathcal{L}$ with typical element ϕ that is based on a set of atoms Atom, where $\top, \bot \in$ Atom and *failure* $\notin$ Atom. Moreover, we define a language $\mathcal{L}_o = \mathcal{L} \cup \{failure\}$ for describing the output of services. We use ϕ not only to denote elements of $\mathcal{L}$, but also of $\mathcal{L}_o$, but if the latter is meant, this will be indicated explicitly. Let N_{sn} with typical element sn be a set of service names such that $d \notin N_{sn}$.

The set of *service descriptions* $\mathcal{S}$ with typical element sd is then defined as follows:

$$\{\langle sn, \mathsf{in}, \mathsf{out}, \mathsf{prec}, \mathsf{eff}\rangle \mid \mathsf{in} \subseteq \mathsf{Atom}, \mathsf{out} \subseteq \mathcal{L}_o, failure \in \mathsf{out}, \mathsf{prec} \in \mathcal{L} \text{ and } \mathsf{eff} \subseteq \mathcal{L}\}.$$

An example from the car repair scenario of a service description is the service that gives information on when it would be possible to make a garage appointment (with some particular garage). In this simple example, the input of this *garageAppInfo* service is {possAppGarageMonday, . . . , possAppGarageFriday}, representing that it can accept information requests regarding whether it would be possible to have an appointment on Monday (*possAppGarageMonday*), Tuesday, etc. The output is the same as the input, with the addition of *failure* and with the addition of the negations of the atoms in the input, representing that it can provide information on possibilities for making an appointment. The idea is that on an input of, e.g., *possAppGarageMonday*, returns either this formula itself or the negation, depending on whether it is possible to make a garage appointment on Monday. Both the preconditions and effects of the service are empty (i.e., $\top$), expressing that the service does not make any changes in the world.

In order to provide some more guidance and intuition to the use of service descriptions, we refine the above definition such that two kinds of services can be described: *information providing* services and *world altering* services [15]. Intuitively, information providing services such as flight information providers give information on something that is the case in the world, and world altering services such as flight booking services can make changes in the world.

In the context of this paper, we define information providing services as services that have no effect, and for each formula appearing in the output description, the negation of this formula should also be in the output description. The idea here is that the service should be able to provide information on each formula in the output description, meaning that it should be able to tell for each formula ϕ in the output description whether this formula holds in the world. The service should thus be able to return on an input ϕ, either ϕ or $\neg\phi$, depending on whether ϕ holds or does not hold, respectively, and therefore both ϕ and $\neg\phi$ should appear in the output description.

World altering services are here defined as services for which the output description is equal to the effects description. The idea is that a world altering service should be able to return the effect of its execution to its caller. As will be explained in more detail in Section 3.1, in order to make full use of goal-based orchestration, it is important that a service returns what it has done. The two kinds of services are defined formally below.

Definition 2 *(information providing and world altering services)*. Let $\langle sn, \mathsf{in}, \mathsf{out}, \mathsf{prec}, \mathsf{eff} \rangle$ be a service description. This service description is an information providing service iff $\mathsf{eff} \equiv \top$ and for each $\phi \neq failure \in \mathsf{out}$, there is a $\phi' \in \mathsf{out}$ such that $\phi' \equiv \neg\phi$. The service description is a world altering service iff $\mathsf{out} \setminus \{failure\} \equiv \mathsf{eff}$, i.e., if $\phi \in \mathsf{out} \setminus \{failure\}$, then $\exists\phi' \in \mathsf{eff} : \phi' \equiv \phi$, and vice versa.

A typical example of an information providing service is the *garageAppInfo* service that was mentioned above. The corresponding world altering service for actually making garage appointments, i.e., the *garageAppMaker* service, takes as input {appGarageMonday, . . . , appGarageFriday}, representing that it accepts request for making appointments on Monday, Tuesday, etc. The output description is the same as the input description, with the addition of *failure*, and the effect description is the output description without *failure* (which is in this case equal to the input description), in accordance with the definition of a world altering service. The idea is that the service receives as input, e.g., *appGarage-Monday*, expressing a request for making a garage appointment on Monday. Assuming that this is possible, the world is changed in such a way that the appointment is actually made, and the appointment itself is returned, to let the requester know that it has made the appointment.

In this paper, we assume that a service description describes either an information providing service, or a world altering service. In principle, one could imagine that there are services that are both information providing and world altering. However, we think that distinguishing these two kinds of services provides a conceptually clear guidance for how service descriptions can be used. Moreover, the two kinds of services fit well with the two kinds of goals that we consider here, i.e., test goals and achievement goals (see Section 2.2).

2.2   Orchestration Language

In this section, we describe the agent-based orchestration language. An agent has a *belief base* and a *goal base*. In [20], the belief base is generally a set of propositional formulas. Here, the belief base, typically denoted by σ, is a pair (σ_a, σ_b) of sets of propositional formulas that are mutually consistent. The idea is that σ_b forms the *background* knowledge of the agent, which does not change during execution. The set σ_a forms the *actual* beliefs, which *are* modified during execution. The sets σ_a and σ_b correspond loosely with the A-Box and T-Box of description logics, respectively. The reason that we make this distinction is explained in Section 3.2.

The goals of the agent can be of two kinds, i.e., a goal is either an *achievement goal* or a *test goal*. An achievement goal $!\phi$, where ϕ is a propositional formula, represents that the agent wants to achieve a situation in which ϕ holds. A test goal $?\phi$ represents that the agent wants to know whether ϕ holds.[1] The idea is that test goals are to be fulfilled by information providing services, and achievement goals may be fulfilled by world altering services. Belief bases and goal bases are defined formally below.

Definition 3 *(belief base and goal base).* The set of belief bases Σ with typical element σ is defined as $\{(\sigma_a, \sigma_b) \mid \sigma_a, \sigma_b \subseteq \mathcal{L}, \ \sigma_a \cup \sigma_b \not\models \bot\}$. The set of goals $\mathcal{L}_\mathsf{G}$ with typical element κ is defined as $\{?\phi, !\phi \mid \phi \in \mathcal{L}\}$. A goal base γ is a subset of $\mathcal{L}_\mathsf{G}$, i.e., $\gamma \subseteq \mathcal{L}_\mathsf{G}$.

In the car repair scenario, the initial goal base contains *!carRepaired*, representing that the agent has the goal of getting the car repaired. The part of the initial belief base with actual beliefs contains *repOnSpot*, representing that initially it is assumed that the car is repairable on the spot, and the background knowledge consists of {appGarageMonday → appGarage,...}, expressing that if the agent has a garage appointment on Monday etc., it has a garage appointment.

Following agent terminology, an expression in our orchestration language is called a *plan*. Broadly speaking, a plan consists of actions which change the belief base of the agent, of subgoals that are to be achieved by selecting a more concrete plan, and of service calls, the results of which may be passed along and which may be stored in the belief base.

A service call has the form $sn^r(\phi, \kappa)$, where sn is the name of the service that is to be called, κ represents the goal that is to be achieved through calling the service, and ϕ represents additional information that forms input to the service. The parameter r is called the revision parameter. This revision parameter can be either np or p, where np stands for non-persistent (meaning that the result of the service call is not stored), and p stands for persistent (meaning that the result is stored in the belief base). We thus provide the programmer with the possibility to specify what to do with the result of service calls. Typically, the results of world altering service calls will be stored in the belief base, together with results of information providing services that are likely to be needed at a later stage.

---

[1] The syntax of representing achievement goals and test goals is taken from [18].

The service name sn of an annotated service call may be either from the set of service names N_{sn}, or it may be the reserved name d. If a name $sn \in N_{sn}$ is used in a service call, we say that this is a *service call by name*[2], i.e., the service with name sn should be called. Usage of the name d in a service call represents a *service call by discovery*, i.e., a service should be discovered that matches the goal with which the service is called.

Sequential composition of actions and service calls is done by means of the construct $b >x> \pi$, where b is an action, a subgoal, or a service call. The result returned from b is bound to the variable x, which may be used in the remaining plan π. This construct for sequential composition is inspired by a similar construct in the orchestration language Orc [7]. Note that the result of a service call can thus be used in the remaining plan, even though it is not stored in the belief base.

Definition 4 *(plan)*. Assume that a set BasicAction with typical element a, and a set of variable names Var with typical element x are given. Let N_{sn}^+ be defined as $N_{sn} \cup \{d\}$.[3] Let $sn \in N_{sn}^+$, $r \in \{np, p\}$, $\phi \in \mathcal{L}$ and $\kappa \in \mathcal{L}_G$. Then the set of plans Plan with typical element π is defined as follows, where b stands for basic plan element.

$$b ::= a \mid \kappa \mid sn^r(\phi, \kappa)$$
$$\pi ::= b \mid b >x> \pi$$

As in Orc, a plan of the form $b \gg \pi$ is used to abbreviate a plan $b >x> \pi$ where x does not occur in π. This may be used in particular in case b is a basic action, as the execution of an action only modifies the belief base and does not return a result. Extending the syntax of plans with more involved constructs, such as constructs for programming parallelism, is left for future research.

An example of a plan in the car repair scenario is (where we abbreviate "Monday" with "M", etc., and if the information parameter of a service call is not shown, this should be interpreted as being $\top$):

$$d^{np}(?(\text{possAppGarageM} \vee \ldots \vee \text{possAppGarageF})) >poss>$$
$$\text{chooseApp}^{np}(poss, ?(\text{appGarageM} \vee \ldots \vee \text{appGarageF})) >app> d^p(!app). \quad (1)$$

This plan intuitively represents that a service should be discovered that provides information on when a garage appointment would be possible, e.g., on Monday and on Tuesday. In our example, the service to be discovered would be the *garageAppInfo* service, as described in Section 2.1. The idea is that one does not know beforehand where a car breakdown will occur, and therefore the orchestration expresses that a service for making garage appointments should be discovered. The result of the service is passed to a service that chooses an

---

[2] Note that usage of the term *call by name* here is not related to the distinction between call by name and call by value in programming language research.

[3] We use sn as typical element of N_{sn} and of N_{sn}^+. It will generally be clear from the context which is meant, and otherwise it will be indicated explicitly.

appointment from possible appointments.[4] The result of the *chooseApp* service, e.g., *appGarageMonday*, is passed to a service for making garage appointments, which needs to be discovered.[5] The intermediate results of the first two service calls are passed along and not stored anywhere, and the result of the service that actually makes the appointment *is* stored.

Plans are executed in order to achieve the agent's goals. The specification of which plan may be executed in order to achieve a certain goal is done by means of *plan selection rules* [20]. A plan selection rule $\kappa \mid \beta \Rightarrow \pi$ intuitively represents that if the agent has the goal κ and believes β to be the case, it may execute the plan π.

Definition 5 *(plan selection rules).* The set of plan selection rules $\mathcal{R}_{\mathsf{PS}}$ is defined as $\{\kappa \mid \beta \Rightarrow \pi \ : \ \kappa \in \mathcal{L}_{\mathsf{G}}, \ \beta \in \mathcal{L}, \ \pi \in \mathsf{Plan}\}$[6].

Plan selection rules can be applied to an agent's top-level goals in the goal base, but also to (the goals of) service calls in a plan that is currently executing (if the service call has not yielded a satisfactory result). Our example agent has two plan selection rules that specify how the goal of getting the car repaired can be reached. The first rule, which we omit here, specifies that a road assistance company can be called if the car is believed to be repairable on the spot. As we assume initially that the agent beliefs the car to be repairable on the spot, this rule is applied first. If it turns our that the car is after all not repairable on the spot, then the second rule can be applied:

$$!\text{carRepaired} \mid \neg\text{repOnSpot} \Rightarrow$$
$$!\text{appGarage} \gg d^p(!\text{appTowTruck}) \gg \text{monitor}^p(?\text{carRepaired}). \quad (2)$$

This rule says that if the agent has the goal of getting his car repaired and he believes it is not possible to repair the car on the spot, it should make a garage appointment and a tow truck appointment, and then it should check whether the car is actually repaired. In order to achieve the goal of having a garage appointment, the agent can apply the plan selection rule $!\text{appGarage} \mid \top \Rightarrow \pi$, where π is the plan from (1).

The mechanism of applying plan selection rules is formalized using the notion of a stack. This stack can be compared with the stack resulting from procedure calls in procedural programming, or method calls in object-oriented programming, and a similar mechanism was also used in [21]. During execution of the agent, a single stack is built. Each element of the stack represents, broadly

---

[4] We assume that the *chooseApp* service returns just one possible appointment from the possible appointments.

[5] Presumably, this should be a service of the same garage as the discovered service for providing information on possible appointments. Extending the orchestration language with a linguistic mechanism for expressing this (using service variables), is left for future research.

[6] We use the notation $\{\ldots \ : \ \ldots\}$ instead of $\{\ldots \mid \ldots\}$ to define sets, to prevent confusing usage of the symbol $\mid$ in this definition.

speaking, the application of plan selection rules to a particular (sub)goal. To be more specific, each element of the stack is of the form $(\pi, \kappa, \mathsf{PS})$, where κ is the (sub)goal to which the plan selection rule has been applied, π is the plan currently being executed in order to achieve κ, and PS is the set of plan selection rules that have not yet been tried in order to achieve κ.

Definition 6 *(stack).* The set of stacks Stack with typical element St to denote arbitrary stacks, and st to denote single elements of a stack, is defined as follows, where $\pi \in \mathsf{Plan}$, $\kappa \in \mathcal{L}_\mathsf{G}$, and $\mathsf{PS} \subseteq \mathcal{R}_\mathsf{PS}$.

$$st ::= (\pi, \kappa, \mathsf{PS})$$
$$St ::= st \mid st.St$$

E is used to denote the empty stack (or the empty stack element), and $E.St$ is identified with St.

We are now in a position to give a definition of an agent. An agent has a belief base, a goal base, a stack, a set of plan selection rules, and a belief update function. The belief update function is introduced as usual [20] for technical convenience, and is used to define the semantics of action execution. We introduce a constraint on agents that expresses that any goal that is used should be consistent with the background knowledge of the belief base. Note that according to this definition, in particular goals $?\top$, $?\bot$, and $!\bot$ are not allowed.

Definition 7 *(agent).* An agent $\mathcal{A}$ is a tuple $\langle \sigma, \gamma, St, \mathsf{PS}, \mathcal{T} \rangle$ where $\sigma \in \Sigma$ is the belief base, $\gamma \subseteq \mathcal{L}_\mathsf{G}$ is the goal base, $St \in \mathsf{Stack}$ is the current plan stack of the agent, $\mathsf{PS} \subseteq \mathcal{R}_\mathsf{PS}$ is a finite set of plan selection rules, and $\mathcal{T}$ is a partial function of type $(\mathsf{BasicAction} \times \Sigma) \rightarrow \Sigma$ and specifies the belief update resulting from the execution of basic actions.

Further, agents should satisfy the following constraint. Let $\sigma = (\sigma_a, \sigma_b)$. It should then be the case that for any goal $\cdot\phi$, where "$\cdot$" stands for $?$ or $!$, occurring in a goal in γ or in a service call in one of the plans of PS, $\sigma_b \not\models \neg\phi$. For any goal $?\phi$ it should also be the case that $\sigma_b \not\models \phi$. Initially, the plan stack of the agent is empty.

3   Semantics

In this section, we define the semantics of the orchestration language. The definition is split into two parts. First, we define the semantics of service calls (Section 3.1), and then we define the semantics of the orchestration language as a whole, making use of the semantics of service calls (Section 3.2).

3.1   Service Calls

In defining the semantics of service calls, we have to define two things. First, we need a definition of when a service matches a service call. Second, we need to specify what a service may return, if it is called.

Although it is not the purpose of this paper to define advanced matching algorithms, we do provide one possible definition of matching. The reason for this is that in this goal-oriented context, the definition of a match depends on the (kind of) goal with which a service is called. We think it is important to identify how the use of goals influences the definition of a match, in order to be able to identify at a later stage which existing matching algorithms can be used in a goal-oriented setting, or how they might have to be adapted.

When matching a service to a goal, the idea is that a test goal is matched to an information providing service, and an achievement goal is matched to a world altering service. That is, for a test goal it is important to match the goal against the *output* description, and for an achievement goal the goal is matched to the *effect* description. The matching definition below corresponds loosely with what is called plug-in matching in [11].

This approach to matching is based on the idea that a service should provide *at least* the information that is asked for, or do *at least* what is desired. Formally, a service description sd matches a service call $sn(\phi, ?\phi')$ if ϕ and ϕ' do not contain atoms that are not in the inputs description of sd. Moreover, what the agent believes to be the case should not contradict with the preconditions description of sd. The idea here is that the agent may not always be able to check whether the precondition of a service holds, but it should at least not have explicit information that the precondition does *not* hold. Finally, there should be a consistent subset of the outputs description (for test goals) or effects description (for achievement goals) from which the goal of the service call follows. Intuitively, this represents that the service is able to provide at least the information that is asked for, or is able to do at least what is desired, respectively.

Definition 8 *(matching a service to a goal).* In the following, we define $\sigma \models \phi$ where $\sigma = (\sigma_a, \sigma_b)$ as $\sigma_a \cup \sigma_b \models \phi$, where $\models$ is the standard entailment relation of propositional logic. Assume a function $atoms : \mathcal{L} \rightarrow \wp(\mathsf{Atom})$ that takes a formula from $\mathcal{L}$ and yields the set of atoms that occur in the formula. Let $sd = \langle sn', \mathsf{in}, \mathsf{out}, \mathsf{prec}, \mathsf{eff} \rangle$ be a service description. Then the matching predicate $\mathrm{match}(sn(\phi, \kappa), \sigma, sd)$, which takes a service call $sn(\phi, \kappa)$, a belief base σ, and a service description sd, is defined as follows if $sn \neq d$.

$$
\begin{aligned}
\mathrm{match}(sn(\phi, ?\phi'), \sigma, sd) \Leftrightarrow\ & sd \text{ is information providing and } sn = sn' \text{ and} \\
& atoms(\phi), atoms(\phi') \subseteq \mathsf{in} \text{ and } \sigma \not\models \neg\mathsf{prec} \text{ and} \\
& \exists \mathsf{out}' \subseteq \mathsf{out} : \mathsf{out}' \not\models \bot \text{ and } \mathsf{out}' \models \phi' \\
\mathrm{match}(sn(\phi, !\phi'), \sigma, sd) \Leftrightarrow\ & sd \text{ is world altering and } sn = sn' \text{ and} \\
& atoms(\phi), atoms(\phi') \subseteq \mathsf{in} \text{ and } \sigma \not\models \neg\mathsf{prec} \text{ and} \\
& \exists \mathsf{eff}' \subseteq \mathsf{eff} : \mathsf{eff}' \not\models \bot \text{ and } \mathsf{eff}' \models \phi'
\end{aligned}
$$

If $sn = d$, then the same definition applies, but the requirement that $sn = sn'$ is dropped.

Note that one needs to define a match by specifying that a goal is a logical consequence of a *consistent subset*, rather than as a logical consequence of the outputs or effects description as a whole, as these descriptions may be inconsistent. This definition is inspired by the so-called consistent subset semantics as proposed in [20, Chapter 4] for defining semantics of goals in case these goals may be inconsistent. Also, note that a service call by name has the additional restriction that the name of the service call should match the name of the service description, meaning that a service call by name is more restrictive than a service call by discovery. Further, if a service is able to provide information on ϕ, it can also provide information on $\neg\phi$, as expressed by the following proposition.

Proposition 1

$$\exists out' \subseteq out : out' \not\models \bot \text{ and } out' \models \phi' \Rightarrow \exists out' \subseteq out : out' \not\models \bot \text{ and } out' \models \neg\phi'$$

The semantics of service execution is defined using a predicate $\text{ret}(sd, \phi)$, which specifies that ϕ may be returned by the service corresponding with service description sd. The idea is that what is returned by a service should be compatible with its output description. If the predicate $\text{ret}(sd, \phi)$ is true, this represents that ϕ may be returned by a service that has service description sd. It is important to have a specification of what may be returned by the service, as we will need it in the semantics of the orchestration language to determine whether the goal of a service call is reached.

Definition 9 *(semantics of service execution).* Let $sd = \langle sn, \text{in}, \text{out}, \text{prec}, \text{eff} \rangle$ be a service description. The predicate *ret* is then defined as follows.

$$\text{ret}(sd, \phi) \Leftrightarrow \phi \equiv \textit{failure} \text{ or}$$
$$\exists out' \subseteq out \setminus \{\textit{failure}\} : (out' \not\models \bot \text{ and } \bigwedge_{\phi_o \in out'} \phi_o \equiv \phi)$$

3.2   Orchestration Language

The semantics of the orchestration language is defined by means of a transition system [17]. A transition system for a programming language consists of a set of axioms and transition rules for deriving transitions for this language. A transition is a transformation of one configuration (or agent in this case) into another and it corresponds to a single computation step.

For reasons of presentation, we will in the following omit the set of plan selection rules PS and the function $\mathcal{T}$ from the specification of an agent, as these do not change during computation. In the transition rules below, we will refer to the set of plan selection rules of the agent with PS_A. Further, we assume the agent has access to a finite set of service descriptions S_A. Finally, we sometimes omit the revision parameter r from service calls, if this parameter is not relevant there.

The first transition rule specifies how a transition for a composed stack can be derived, given a transition for a single stack element. It specifies that only the top element of a stack can be transformed or executed.[7]

---

[7] We omit a rule that specifies that the two topmost stack elements may be modified at the same time.

Definition 10 *(stack execution).* Let $st \neq E$.

$$\frac{\langle \sigma, \gamma, \mathrm{st} \rangle \rightarrow \langle \sigma', \gamma', \mathrm{st}' \rangle}{\langle \sigma, \gamma, \mathrm{st.St} \rangle \rightarrow \langle \sigma', \gamma', \mathrm{st'.St} \rangle}$$

Before we continue with the definition of transition rules, we need to define when a goal is achieved, and when a plan selection rule can be applied to a goal. The definition of when a goal is achieved differs for test goals and achievement goals. A test goal is evaluated against the result of a service call, i.e., the belief base is not taken into account. The idea is that a test goal is used if the agent wants to obtain or to check a piece of information, regardless of whether it already believes something about this piece of information. The achievement of an achievement goal, on the other hand, is determined on the basis of the belief base, together with the result of a service call.

We specify what it means to take a belief base together with a result of service execution, using a belief revision function. For examples on how such a function is defined, see, e.g., [1] which shows how a belief revision mechanism can be incorporated into the agent programming language AgentSpeak(L). Below, we only specify the constraints that such a belief revision function should satisfy. That is, if a belief base (σ_a, σ_b) is updated with a result x, only σ_a should be updated. The function is not defined if x is inconsistent with the background knowledge.

Definition 11 *(belief revision function).* In the following, we assume a partial belief revision function $brev : (\wp(\mathcal{L}) \times \wp(\mathcal{L})) \rightarrow (\mathcal{L} \rightarrow \wp(\mathcal{L}))$. The function $brev((\sigma_a, \sigma_b), x)$ is defined iff $\sigma_b \not\models \neg x$, and if it is defined, it should satisfy the following constraints on behavior: $brev((\sigma_a, \sigma_b), x) = (\sigma'_a, \sigma_b)$ where $\sigma'_a \models x$ and $\sigma'_a \cup \sigma_b \not\models \bot$; if $\sigma \models x$, then $brev(\sigma, x) = \sigma$.

In agent programming frameworks, the achievement of achievement goals is determined on the belief base, as this is the only component representing the current situation. In this context where we have service calls that return results, not all results are stored in the belief base. Therefore, we also take into account the result of the relevant service call when evaluating whether a goal is achieved. In fact, the evaluation of a test goal should be performed *only* on the result of a service call, as we disregard whether the agent already believes something about the test goal.

This is reflected in the semantics of goal achievement as defined formally below, in which we specify when a predicate $\mathrm{ach}(\kappa, \sigma, x)$ holds, representing that the goal κ is achieved with respect to the result of a service call x and belief base σ. To be more specific, a test goal $?\phi$ holds if the result of the service call expresses that either ϕ or $\neg\phi$ hold. An achievement goal $!\phi$ holds if ϕ follows from the belief base that would result from updating the old belief base with the service result.

Definition 12 *(semantics of goal achievement).* The semantics of goal achievement is defined as a predicate $\mathrm{ach}(\kappa, \sigma, x)$ that takes a goal κ, a belief base σ, and a propositional formula $x \in \mathcal{L}$ that represents the result against which κ should be checked.

$$\text{ach}(?\phi, \sigma, x) \Leftrightarrow brev(\sigma, x) = \sigma' \text{ and } (x \models \phi \text{ or } x \models \neg\phi) \text{ and } x \neq failure$$
$$\text{ach}(!\phi, \sigma, x) \Leftrightarrow brev(\sigma, x) = \sigma' \text{ and } \sigma' \models \phi \text{ and } x \neq failure$$

Although test goal achievement is defined in principle only on the result of the service call x, we use the belief base to check that x is consistent with the background knowledge. We thus have by definition that a goal cannot be achieved with respect to a result of a service call, if this result is inconsistent with the background knowledge (in that case $brev$ would be undefined for this result).

A plan selection rule ρ of the form $\kappa' \mid \beta \Rightarrow \pi$ is applicable to a goal κ given a belief base σ, if κ' matches κ, β holds according to σ, and κ' is not achieved. The kind of matching we use is one that could be called *partial matching*. Here, a rule with head κ' is applicable to a goal κ if κ' "follows from" κ. That is, a rule with, e.g., head $!p$, could match a goal $!(p \wedge q)$. This kind of semantics is generally used for these rules in agent programming [20], as the goal decomposition for which it allows has practical advantages.

Definition 13 *(applicability of plan selection rule)*. We define a predicate applicable(ρ, κ, σ) that takes a plan selection rule ρ, a goal κ, and a belief base σ as follows, where "·" stands for ? or !.

$$\text{applicable}(\cdot\, \phi' \mid \beta \Rightarrow \pi, \cdot\, \phi, \sigma) \Leftrightarrow \phi \models \phi' \text{ and } \sigma \models \beta \text{ and } \neg\text{ach}(\cdot\, \phi', \sigma, \top)$$

In order to start execution and create a first stack element, the agent applies a plan selection rule to a goal in the goal base. However, we leave out the corresponding transition rule for reasons of space.

When an agent encounters a service call construct during execution of a plan, it tries to call matching services until there are no more matching services, or the goal of the service call is reached. In order to keep track of which services have been called, we annotate the service call construct (initially) with the set of services that are available. For an achievement goal, the agent only tries to call services if the goal is not already reached. For a test goal, the agent always tries to call a service, no matter whether it already believes the test goal to hold. The service call is then used to check whether the information of the agent is correct. We omit the rule for test goals for reasons of space. From this set of services, a matching service is selected non-deterministically. The result of the execution of the selected service is also stored in the annotation with the service call construct, as this is used in other transition rules (Definitions 16 and 22) to check whether the goal of the service call is reached. For reasons of presentation, we omit here and in the sequel rules for dealing with service calls and actions that form the last element of a plan.

Definition 14 *(calling services)*

$$\frac{\neg\text{ach}(!\phi', \sigma, \top)}{\langle \sigma, \gamma, (sn(\phi, !\phi') >x> \pi, \kappa, \text{PS})\rangle \rightarrow \langle \sigma, \gamma, (sn(\phi, !\phi')[S_{\mathcal{A}}, \top] >x> \pi, \kappa, \text{PS})\rangle}$$

$$\frac{\neg\mathrm{ach}(\kappa,\sigma,x_o) \quad sd \in S \quad \mathrm{match}(sn(\phi,\kappa),\sigma,sd) \quad \mathrm{ret}(sd,x_n)}{\langle\sigma,\gamma,(sn(\phi,\kappa)[S,x_o] >x> \pi,\kappa',\mathsf{PS})\rangle \rightarrow \\ \langle\sigma,\gamma,(sn(\phi,\kappa)[S \setminus \{sd\},x_n] >x> \pi,\kappa',\mathsf{PS})\rangle}$$

Note that the second transition rule selects a matching service if the goal of the service call is not reached. This provides a way of dealing with failure of services, as another service is tried if a former service call did not have the desired result. This is easily specified in our semantics, as we use an explicit representation of goals.

In the following, we use a revision function that takes a revision parameter r, a belief base σ and a result of a service call x, and updates σ with x, depending on r.

Definition 15 *(revision function)*. The revision function rev is defined as follows: $rev(np,\sigma,x) = \sigma$ and $rev(p,\sigma,x) = brev(\sigma,x)$.

The next transition rule specifies what happens if the goal of a service call is reached after calling a service, and the goal of the stack element is not yet reached. If this is the case, the belief base is updated according to the revision parameter, and all occurrences of the parameter x of the sequential composition in the rest of the plan π are replaced by the result of the service call x', i.e., the result of the service call is passed along. Moreover, all goals in the goal base that are believed to be reached after the revision resulting from the service call are removed.

Definition 16 *(goal of service call achieved after service execution)*

$$\frac{\neg\mathrm{ach}(\kappa',\sigma,x') \quad \mathrm{ach}(\kappa,\sigma,x') \quad rev(r,\sigma,x') = \sigma' \quad \gamma' = \gamma \setminus \{\kappa \mid \mathrm{ach}(\kappa,\sigma,x')\}}{\langle\sigma,\gamma,(sn^r(\phi,\kappa)[S,x'] >x> \pi,\kappa',\mathsf{PS})\rangle \rightarrow \langle\sigma',\gamma',([x'/x]\pi,\kappa',\mathsf{PS})\rangle}$$

It might also be possible that a subgoal or the goal of a service call is already reached before a service is called (only in case of an achievement goal). The question is, what should be passed along in this case. One possibility would be to pass along the goal itself. However, this yields unintuitive results in case the background knowledge is used to derive the goal. In the car repair scenario, the subgoal !appGarage can be achieved by applying a plan selection rule !appGarage | $\top \Rightarrow \pi$, where π is the plan from (1). The goal !appGarage can be achieved through the service call $d^p(!app)$, which is matched to the *garageApp-Maker* service. This service might return, e.g., *appGarageMonday*. Taking the background knowledge of the agent, we can then derive !appGarage, making this goal achieved.

The idea now is, that we want to pass along *appGarageMonday*, rather than *appGarage*, as the first is the concrete realization of the second higher level goal. This is achieved by passing along only the σ_a part of the belief base that "contributes" to the goal being reached. That is, background knowledge is not passed along. The part of σ_a that should be passed along, is what we call the *base*, and this is defined formally below.

Definition 17 *(base)*. The predicate $\mathrm{base}(\sigma, \phi, x)$ takes a belief base σ, a formula ϕ where $\sigma \models \phi$, and a formula x representing the base of ϕ in σ. Let $\sigma = (\sigma_a, \sigma_b)$, and let $\sigma'_a \subseteq \sigma_a$ such that $(\sigma'_a, \sigma_b) \models \phi$ and for any σ''_a such that $\sigma''_a \subset \sigma'_a$, we have $(\sigma''_a, \sigma_b) \not\models \phi$. The predicate is then defined as follows: $\mathrm{base}((\sigma_a, \sigma_b), \phi, x) \Leftrightarrow x = \bigwedge_{\phi' \in \sigma'_a} \phi'$.

The idea of the base is thus to extract the concrete realization of a goal, rather than the higher level abstract goal. In description logic, a concrete realization might correspond with an instance, where the higher level goal could be represented using a concept. The transition rule below specifies the case where the goal is reached before a service is called. Is similar rule is used for subgoals, but we omit it for reasons of space.

Definition 18 *(goal of service call achieved before services are called)*

$$\frac{\neg\mathrm{ach}(\kappa', \sigma, \top) \quad \mathrm{ach}(!\phi', \sigma, \top) \quad \mathrm{base}(\sigma, \phi', x')}{\langle \sigma, \gamma, (sn(\phi, !\phi') >x> \pi, \kappa', \mathsf{PS}) \rangle \to \langle \sigma, \gamma, ([x'/x]\pi, \kappa', \mathsf{PS}) \rangle}$$

If a subgoal is not achieved, a plan selection rule may be applied to the subgoal. The application of a plan selection rule to a (sub)goal is the only way in which a new stack element can be created.

Definition 19 *(apply rule to create stack element)*. Below, $\mathsf{PS}' = \mathsf{PS}_{\mathcal{A}} \setminus \{\kappa' \mid \beta \Rightarrow \pi\}$.

$$\frac{\neg\mathrm{ach}(\kappa'', \sigma, \top) \quad \kappa' \mid \beta \Rightarrow \pi \in \mathsf{PS}_{\mathcal{A}} \quad \mathrm{applicable}(\kappa' \mid \beta \Rightarrow \pi, \kappa, \sigma)}{\langle \sigma, \gamma, (\kappa >x> \pi', \kappa'', \mathsf{PS}) \rangle \to \langle \sigma, \gamma, (\pi, \kappa, \mathsf{PS}').(\kappa >x> \pi', \kappa'', \mathsf{PS}) \rangle}$$

If the plan of a stack element is empty, a plan selection rule may be applied in order to select another plan to try to reach the goal of the stack element. Note that if a plan selection rule is applied to the goal of a stack element, this does not lead to the creation of a new stack element.

Definition 20 *(apply rule within stack element)*. Below, $\mathsf{PS}' = \mathsf{PS} \setminus \{\kappa' \mid \beta \Rightarrow \pi\}$.

$$\frac{\kappa' \mid \beta \Rightarrow \pi \in \mathsf{PS} \quad \mathrm{applicable}(\kappa' \mid \beta \Rightarrow \pi, \kappa, \sigma)}{\langle \sigma, \gamma, (\epsilon, \kappa, \mathsf{PS}) \rangle \to \langle \sigma, \gamma, (\pi, \kappa, \mathsf{PS}') \rangle}$$

Popping a stack element is done in two cases: the goal of the stack element is reached, or the goal is not reached and there are no more applicable rules. The goal of a stack element may be reached after a service call, or after action execution. In the first case, the result of the relevant service call is passed to the stack element just below the top element. In the second case, a result to be passed is obtained from the belief base using the *base* predicate (Definition 17).

Definition 21 *(popping a stack element: goal of stack element reached or unreachable)*

$$\frac{\mathrm{ach}(\kappa_1, \sigma, x) \quad \mathrm{rev}(r, \sigma, x) = \sigma' \quad \gamma' = \gamma \setminus \{\kappa \mid \mathrm{ach}(\kappa, \sigma', \top)\}}{\langle \sigma, \gamma, (sn^r_1(\phi_1, \kappa_3)[S, x] >x_1> \pi_1, \kappa_1, \mathsf{PS}_1).(\kappa_1 >x_2> \pi_2, \kappa_2, \mathsf{PS}_2) \rangle \to}$$
$$\langle \sigma', \gamma', ([x/x_2]\pi_2, \kappa_2, \mathsf{PS}_2) \rangle$$

$$\frac{\mathcal{T}(\sigma, a) = \sigma' \quad \text{ach}(!\phi', \sigma', \top) \quad \text{base}(\sigma', !\phi', x') \quad \gamma' = \gamma \setminus \{\kappa \mid \text{ach}(\kappa, \sigma', \top)\}}{\langle \sigma, \gamma, (a \gg \pi', !\phi', \mathsf{PS'}).(!\phi' > x > \pi, \kappa, \mathsf{PS}) \rangle \rightarrow \langle \sigma', \gamma', ([x'/x]\pi, \kappa, \mathsf{PS}) \rangle}$$

$$\frac{\neg \text{ach}(\kappa, \sigma, \top) \quad \neg \exists \rho \in \mathsf{PS} : \text{applicable}(\rho, \kappa, \sigma)}{\langle \sigma, \gamma, (\epsilon, \kappa, \mathsf{PS}).(\kappa > x > \pi, \kappa', \mathsf{PS'}) \rangle \rightarrow \langle \sigma, \gamma, (\epsilon, \kappa', \mathsf{PS'}) \rangle}$$

Note that if the plan of a stack element is empty, the goal of the stack element is not reached. The reason is that the stack element would have been popped before the plan got empty, if the goal of stack element would have been reached after a service call or an action execution.

If an action of a plan cannot be executed, or there is no applicable rule for a subgoal of a plan, or the goal of a service call has not been reached and there are no more services that match, then the plan fails. If this happens, the plan is dropped. Consequently, another plan for achieving the goal of the stack element may be tried, providing for flexibility in handling failure (Definition 20). Below, we only show the transition rule for the case where an action is not executable.

Definition 22 *(plan failure)*

$$\frac{\mathcal{T}(\sigma, a) \text{ is undefined}}{\langle \sigma, \gamma, (a \gg \pi, \kappa, \mathsf{PS}) \rangle \rightarrow \langle \sigma, \gamma, (\epsilon, \kappa, \mathsf{PS}) \rangle}$$

Popping a stack element if the goal is not reached and there are no more applicable rules, prevents the agent from getting "stuck" or from looping inside a stack element, while not reaching the (sub)goal of the stack element. If stack elements are popped if the goal cannot be reached, the agent can try another plan in the stack element that then becomes the new top element of the stack. This mechanism functions recursively, meaning that if the agent has tried everything without success, it will have an empty stack element again. However, the top-level goal that the agent tried to reach is still in the belief base, at it was probably not reached. The agent can then try another goal, or wait for the circumstances to change for the better, and give it another try later on. If the agent terminates, then either the agent has an empty stack and the goal base is empty, or the agent has an empty stack and there are no applicable rules to the goals in the goal base. These considerations of progress and terminations are formalized in the proposition below.

Proposition 2 *(progress and termination)*. Let $\mathcal{A} = \langle \sigma, \gamma, St \rangle$ be an agent. Then, if $St \neq E$, there is an $\mathcal{A}'$ such that $\mathcal{A} \rightarrow \mathcal{A}'$. Further, on any computation $\mathcal{A}, \mathcal{A}_1, \ldots$ there is an $\mathcal{A}'$ of the form $\langle \sigma', \gamma', E \rangle$. Finally, if there is no $\mathcal{A}'$ such that $\mathcal{A} \rightarrow \mathcal{A}'$, then either $\mathcal{A}$ is of the form $\langle \sigma, \emptyset, E \rangle$, or $\mathcal{A}$ is of the form $\langle \sigma, \gamma, E \rangle$ and there is no ρ and $\kappa \in \gamma$ such that applicable$(\rho, \kappa, \sigma, \top)$.

4   Conclusion

In this paper, we have proposed to use goal-oriented techniques from the field of cognitive agent programming for service orchestration. The advantage of using an explicit representation of goals is the flexibility in handling failure that goals provide. To be more specific, goals provide for flexibility in at least four ways. First, goals can be used to do semantic matchmaking, yielding flexibility in selection of services, as one does not necessarily have to define at design time which particular service should be called. Second, the explicit use of goals makes it easy to check whether a service call was successful, making it easy to build into the semantics a mechanism for trying other matching services if one service fails. Third, the plan selection rules can be used in a natural way to specify an alternative plan if calling a service directly fails. Finally, the possibility to specify *multiple* plans for achieving a goal in combination with the mechanism for detecting whether a goal is achieved, can make the orchestration more flexible in handling failure, and also more responsive to the actual situation as plan selection rules are conditionalized on beliefs. We have made these ideas concrete by formally defining a goal-based orchestration language that makes use of semantic matchmaking.

We see several directions for future research. One of the main issues that needs to be dealt with is the fact that the language is based on propositional logic. Important candidates to replace propositional logic are description logics, given the fact that we incorporate semantic matchmaking into our framework, or first-order logics. We plan to study in particular whether the WSML [8] language can be integrated into our work, as this is a language that has both first-order logic and description logic variants. Further, WSML seems to be particularly suited for our work, as it provides a formal syntax and semantics for WSMO, which is based on goals. Moreover, we plan to extend the language of plans to include more sophisticated constructs such as a construct for parallel composition. Further, we consider to investigate more expressive kinds of service communication in which interaction protocols are used for communication. Furthermore, we want to investigate how work on soft constraints [3] can be used to obtain a more expressive language for representing goals and for defining goal achievement. Moreover, we want to analyze how exactly our semantics for the result passing sequential composition construct differs from the Orc semantics. Finally, we aim to analyze formally how our goal-based mechanism for handling failure is related to more conventional approaches to failure handling, such as used in WS-BPEL.

Concluding, we believe that the framework as laid out in this paper can provide a foundation for several interesting directions of future research, and we hope it contributes to the further investigation of combining agent-oriented and service-oriented approaches.

References

1. N. Alechina, R. Bordini, J. Hübner, M. Jago, and B. Logan. Automating belief revision for agentspeak. In *Proc. of DALT'06*, 2006.
2. M. Baldoni, C. Baroglio, A. Martelli, V. Patti, and C. Schifanella. Interaction protocols and capabilities: A preliminary report. In *Proc. of PPSWR'06*, pages 63–77, 2006.
3. S. Bistarelli, U. Montanari, and F. Rossi. Semiring-based constraint solving and optimization. *Journal of ACM*, 44:201–236, 1997.

4. L. Bozzo, V. Mascardi, D. Ancona, and P. Busetta. CooWS: Adaptive BDI agents meet service-oriented programming. In *Proc. of WWW/Internet'05*, volume 2, pages 205–209. IADIS Press, 2005.
5. L. Braubach, A. Pokahr, D. Moldt, and W. Lamersdorf. Goal representation for BDI agent systems. In *Proc. of ProMAS'04*, volume 3346 of *LNAI*, pages 44–65. Springer, Berlin, 2005.
6. R. Bruni, H. Melgratti, and U. Montanari. Theoretical foundations for compensations in flow composition languages. In *Proc. of POPL'05*, pages 209–220, 2005.
7. W. R. Cook and J. Misra. Computation orchestration: A basis for wide-area computing, 2007. To appear in the Journal on Software and System Modeling.
8. J. de Bruijn, H. Lausen, R. Krummenacher, A. Polleres, L. Predoiu, M. Kifer, and D. Fensel. The web service modeling language WSML. WSML deliverable d16.1v0.2, 2005. `http://www.wsmo.org/TR/d16/d16.1/v0.2/`.
9. I. Dickinson and M. Wooldridge. Agents are not (just) web services: considering BDI agents and web services. In *Proc. of SOCABE'05*, 2005.
10. M. Juric, P. Sarang, and B. Mathew. *Business Process Execution Language for Web Services 2nd Edition*. Packt Publishing, 2006.
11. L. Li and I. Horrocks. A software framework for matchmaking based on semantic web technology. In *Proc. of WWW'03*, pages 331–339. ACM Press, 2003.
12. D. Martin, M. Paolucci, S. McIlraith, M. Burstein, D. McDermott, D. McGuinness, B. Parsia, T. Payne, M. Sabou, M. Solanki, N. Srinivasan, and K. Sycara. Bringing semantics to web services: The OWL-S approach. In *Proc. of SWSWPC 2004*, volume 3387 of *LNCS*, pages 26–42, 2005. Springer, Berlin.
13. V. Mascardi and G. Casella. Intelligent agents that reason about web services: a logic programming approach. In *Proc. of ALPSWS'06*, pages 55–70, 2006.
14. M. Mazzara and R. Lucchi. A framework for generic error handling in business processes. *Electr. Notes Theor. Comput. Sci.*, 105:133–145, 2004.
15. S. A. McIlraith, T. C. Son, and H. Zeng. Semantic web services. *IEEE Intelligent Systems*, 16(2):46–53, 2001.
16. M. Paolucci, T. Kawamura, T. R. Payne, and K. P. Sycara. Semantic matching of web services capabilities. In *Proc. of ISWC'02*, volume 2342 of *LNCS*, pages 333–347. Springer-Verlag, 2002.
17. G. D. Plotkin. A Structural Approach to Operational Semantics. Technical Report DAIMI FN-19, University of Aarhus, 1981.
18. A. S. Rao. AgentSpeak(L): BDI agents speak out in a logical computable language. In *Agents Breaking Away (LNAI 1038)*, pages 42–55. Springer-Verlag, 1996.
19. D. Roman, U. Keller, H. Lausen, J. de Bruijn, R. Lara, M. Stollberg, A. Polleres, C. Feier, C. Bussler, and D. Fensel. Web service modeling ontology. *Applied Ontology*, 1:77–106, 2005.
20. M. B. van Riemsdijk. *Cognitive Agent Programming: A Semantic Approach*. PhD thesis, 2006.
21. M. B. van Riemsdijk, M. Dastani, J.-J. Ch. Meyer, and F. S. de Boer. Goal-oriented modularity in agent programming. In *Proc. of AAMAS'06*, pages 1271–1278, Hakodate, 2006.
22. M. Winikoff, L. Padgham, J. Harland, and J. Thangarajah. Declarative and procedural goals in intelligent agent systems. In *Proc. of KR'02*, 2002.
23. M. Wirsing, A. Clark, S. Gilmore, M. Hölzl, A. Knapp, N. Koch, and A. Schroeder. Semantic-based development of service-oriented systems. In *Proc. of FORTE'06*, volume 4229 of *LNCS*, pages 24–45. Springer-Verlag, 2006.

An Agent-Based Approach to User-Initiated Semantic Service Interconnection

Nicolas Braun, Richard Cissée, and Sahin Albayrak

DAI-Labor, TU Berlin,
Franklinstrasse 28/29, 10587 Berlin, Germany
{nicolas.braun,richard.cissee,sahin.albayrak}@dai-labor.de
http://www.dai-labor.de

Abstract. The Internet has emerged as a dynamic service-centered environment. How to combine several individual services to provide value-added services is an ongoing research challenge, mainly because dynamic service composition cannot be accomplished easily and the users' intentions regarding service interconnection have to be anticipated by the service developer. To overcome the drawbacks of developer-centric service composition, we propose an approach to user-initiated dynamic service interconnection characterized by a high degree of usability. We outline a solution for semantic interconnection of two or more services, which provides information that would not be available by using a single service. Based on various interconnection types, we have developed an architecture for user-initiated semantic service interconnection utilizing the planning and negotiation capabilities of intelligent agents. As proof of concept, we have implemented a Web 2.0-based service suite consisting of various individual entertainment services, which assist users in planning their leisure time.

Keywords: User-Initiated Dynamic Service Interconnection, Intelligent Agents.

1 Introduction

Today, an ever-increasing number of services within various domains (such as office, home, or pastime) is available to users over the Internet. In most cases, these services are not interconnected and the user has to integrate the provided information manually, e.g. by copying and pasting data from one service to another, which reduces the overall usability. An approach to overcome this limitation is the provisioning of a suite of integrated services as a single application. In this case, however, the service developer or provider has to manually combine single services in order to obtain composite services. Service composition has numerous drawbacks: As the number of potential service combinations is quadratic (if two services are combined) or exponential (if more than two services are combined) in the number of services, the process is rather time-consuming. Additionally, the developer may lose sight of potential combinations. While there are some

J. Huang et al. (Eds.): SOCASE 2007, LNCS 4504, pp. 49–62, 2007.

approaches (see Section 5 for examples) supporting developers of composite services, e.g. by providing tools, other problems remain: Service composition cannot easily be accomplished dynamically. Whenever a new service is added, existing services have to be modified, and new composite services have to be developed and deployed. Additionally, the user faces a potentially large number of composite services, each of which he has to familiarize himself with.

Our approach addresses these issues by supporting user-initiated service interconnection. We regard this as a more intuitive approach, as the user receives exactly the information he requires. Instead of having to interact with a large number of composite services via various additional user interfaces, the user interacts with a comparatively small amount of services in the usual manner and receives information based on the interconnection in addition to the results of the single services.

Consider the following two scenarios from the domain of entertainment-related services: In the first scenario, the user utilizes a restaurant finder service RS and a weather forecast service WF. Via RS, the user initiates a search for restaurants in a specific city and obtains a list of results. The user then drags and drops the icon representing WF onto the RS dialog displaying the results. In this case, a useful interconnection result would be to indicate for each restaurant whether it is suitable with regard to the current weather forecast for the respective location (as an example, restaurants with open air areas could be rated higher if the weather is fair).

In the second scenario, the user again utilizes RS, obtaining a list of restaurants, and a movie finder service MS. Via MS, the user initiates a search for movies with specific parameters, and obtains a list of results. He selects one movie from the results and interconnects the two services by dragging and dropping the icon representing this movie onto the RS dialog displaying the results. In this case, a useful interconnection result would be to indicate for each restaurant the distance to the nearest cinema showing the movie (or at least a similar movie, if there is no direct match). At this point, the user can chose the most appropriate interconnection step from a drop-down menu.

This paper describes an agent-based approach that allows these kinds of dynamic user-initiated interconnection of services. Multi-Agent System (MAS) technology in particular allows us to utilize capabilities of agents such as the ability to negotiate and plan the usage of agent services in an intelligent manner. As proof of concept and in order to evaluate the approach, we have implemented the Smart Pastime Assistant (SPA) as a suite of prototypical services from the entertainment domain.

The remainder of this paper is structured as follows: Section 2 describes the general concepts and architectural components of our approach. Section 3 outlines the implementation of the approach. In Section 4, we briefly describe the SPA. Section 5 presents related work. Section 6 concludes the paper with an outlook and outlines further work.

2   Approach

The goal of our approach is to allow services to be interconnected semantically, dynamically within a deployed system, and on-the-fly, i.e. without having to adjust the implementation of the respective services. In this context and throughout the paper, the term "service" refers to a sub-application within a larger context that is usable by human users, generally via a GUI. In contrast, the term "agent service" refers to functionality provided by agents for other agents, which is specified via an ontology-based description of preconditions and results of the agent service.

2.1   Service Interconnection

The interconnection of two services (namely, a source service and a target service) is usually triggered by the user interacting with the system by selecting interconnection objects. These objects contain and represent information available to the user, such as a weather forecast for a specific city, or a set of restaurants. Figure 1 gives an overview of an interconnection process. As a result of the interconnection, the user is presented non-obvious information that is not available by using a single service. Other kinds of actions triggering interconnection are conceivable, such as a service interconnection at regular intervals in order to push new information to the user. The subsequent procedure, however, is always independent of the action triggering the interconnection:

- The source service propagates the source interconnection object to the target service.
- Based on this object and its own internal state, represented by the target interconnection object, the target service determines the interconnection type, and obtains the results of the interconnection via applying concrete rules and/or using agent services. This step is described in detail below.
- If any new rules have been obtained in the previous step, they are added to a rule repository.
- The interconnection results are propagated to the user.

Because the interconnection of two services is not hard-coded into either service, the required operations have to be determined via a rule repository

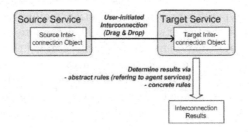

Fig. 1. Overview of an interconnection process

containing rules for interconnecting services (see Section 2.2). There are two kinds of rules: Abstract rules are mapped to agent services, while concrete rules are directly applicable via an agent's rule execution component. Concrete rules are generally preferable because of their lower complexity, which allows them to be applied more efficiently than agent services. On the other hand, agent services are usually more generic and capable of carrying out more complex operations.

Concrete rules may be added to the system either manually by service developers, by users via a GUI, or they may be determined via a reasoner based on observed user interactions. Currently, however, only the first two alternatives are supported by our approach. Concrete rules are either applicable globally or pertain to specific users or user groups. For purposes of planning and aggregating single rules, both types of rules are equivalent, because they may be represented in the same manner: A rule produces, for a list of specific objects as input, a list of specific objects as output. In our approach, rules are specified in the Semantic Web Rule Language (SWRL), which facilitates a direct mapping between rules and the semantic descriptions of our agent services.

The sub-steps of the second step of the interconnection procedure are carried out as follows:

- The target service determines via the rule repository and, if necessary, the planning and aggregation component, whether concrete or abstract rules for the interconnection exist.
- If rules are found, they are applied and the results are returned.
- If no suitable rules are found, the target service attempts to discover further applicable agent services via the MAS infrastructure and via negotiating with potential service providers.
- If agent services are found, they are used in combination with any rules that had been already present and the results are returned. New abstract rules, which refer to the newly discovered agent services, are added to the rule repository.
- If no suitable agent services are found, the target service attempts to obtain the missing rules via direct interaction with the user.
- If rules are provided by the user, they are applied in combination with any rules that had already been present and the results are returned. New concrete rules, as obtained from the user, are added to the rule repository.
- If no suitable rules are provided by the user, the service interconnection fails.

There are various types of service interconnection, mainly based on the current state of the source and the target service, as well as the interconnection objects. An interconnection object may represent a specific state of the respective source or target service, i.e. its current dialog step. In this case, a service icon or an entire service window may represent this object in a GUI-based system. For interconnection purposes, a single dialog step furthermore consists of a set of objects, which may be empty. Thus, one of these objects may also constitute an interconnection object.

Each object of a dialog step is classified either as a query object, in which case it is intended for obtaining information from the user or another external source,

or as a result object, in which case it contains information that is intended to
be presented to the user or another external source. Query objects are usually
required for proceeding to further dialog steps of the respective services, while
result objects constitute a partial or final result of the respective services. A single
dialog step is always considered to primarily contain either query objects or result
objects, although in some cases the distinction is somewhat problematic, mainly
because query objects may be mapped directly to result objects. Whenever an
abstract representation of a service is used as an interconnection object, it is
mapped internally to the set of objects of the respective dialog step. As a single
object from a dialog step is also mapped to a set (of size one), both types of
interconnection objects are subsumed in the following.

In any case, the expected behavior of the target service is to update the
objects of its current dialog step, based on the interconnection results, and to
remain in that dialog step if the target interconnection object is an abstract
representation (unless the service has not been started yet, in which case the
service proceeds to the first dialog step). If the target interconnection object is a
single object, the expected behavior of the service is to proceed to a subsequent
dialog step for presenting the interconnection results, much in the same way as
if the user had selected the respective object itself in the target service without
interconnection. Based on these definitions, we are now able to describe the
different types of service interconnection, which are summarized in Table 1.

Table 1. Overview of service interconnection types. Interconnection by linking two
sets of query objects is ruled out because it is not expected to lead to viable results.

Source Service:	Target Service:	
	Query Objects	Result Objects
Query Objects	N/A	Combined Rule Composition & Aggregation (see Section 2.1)
Result Objects	Rule Composition (see Section 2.1)	Rule Aggregation (see Section 2.1)

Linking Result Objects to Query Objects. In the most straightforward
case, a set of result objects of a source service is linked to a set of query objects
of a target service. This type of interconnection utilizes a rule composition in
which the output of the source service is used, either directly or via applying
a set of rules, as the input of the target service. Similar interconnections are
already present in existing systems (see e.g. [24]). If the result objects cannot be
mapped directly to query objects, applicable rules have to be found either within
the rule repository or via utilizing the planning component. The interconnection

results are used within the query objects, e.g. in the form of additional search parameters, and they are also used in further dialog steps, e.g. as additional references assigned to result objects, or in further interconnections.

As an example, this kind of service interconnection is applicable in a modification of the first scenario described in the introduction: If a restaurant object is dragged and dropped onto the representation of the weather forecast service, the location of the restaurant may be used as input for the weather forecast request, i.e. the parameters are matched directly, as shown in Figure 2. The interconnection result is a weather forecast request for the given location. If the restaurant object is dropped onto the weather forecast request itself, the parameters are matched in the same manner, and additionally the service executes the request and returns a weather forecast as a result. In any case, as soon as the subsequent interaction step is reached, the result objects (weather forecast and restaurants) are automatically interconnected again, as described in the following section.

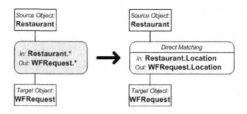

Fig. 2. Interconnection of result and query object via rule composition

Linking Result Objects to Result Objects. In the second type of interconnection, the current dialog step of the target service primarily consists of result objects. In this case, a simple service composition is not applicable because there is no service requiring any input. Therefore, the interconnection is realized as a service aggregation in which the output of both source and target service is aggregated and used as input of a third internal service in order to produce further output. Applicable rules have to be found either within the rule repository or via utilizing the aggregation component. The interconnection results are used in the respective dialog step, e.g. as additional references assigned to the result objects. As an example, this kind of service interconnection is applicable in the second scenario described in the introduction: If a movie object is dragged and dropped onto a restaurant object, a combination of rules has to be determined in order to aggregate these objects as input parameters, as visualized in Figure 3. This type of interconnection is also carried out automatically in dialog steps following a dialog step in which an interconnection process has been carried out.

Linking Query Objects to Result Objects. This type of interconnection is basically a reversion of the first type, combined with the second type: Because the interconnection results are presented within the service providing result objects, a suitable additional result object has to be generated based on the given

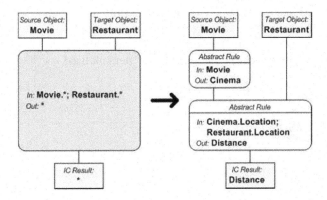

Fig. 3. Interconnection of result objects via rule aggregation

interconnection objects, and this additional result object has to be aggregated with the given result objects. As an example, this kind of service interconnection is applicable in the first scenario described in the introduction: If a representation of the weather forecast service is dragged and dropped onto a restaurant object, the location of the restaurant may again be used as input for the weather forecast request. A weather forecast is obtained as an internal result, and aggregated via concrete rules with the restaurant object, as shown in Figure 4.

2.2 Main Architectural Components

In the previous section, various agents and components have been introduced as required functionality for the various types of service interconnection. Figure 5

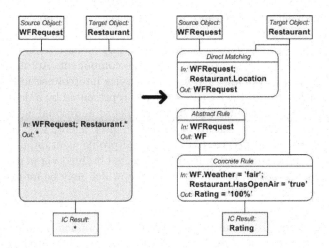

Fig. 4. Interconnection of query and result object via rule composition and aggregation

shows the relationships between these agents and components in the overall architecture. In this paper, we largely omit the description of additional parts of the architecture that are not directly relevant for service interconnection, such as components offering functionality for personalized services, user profile management, and the like.

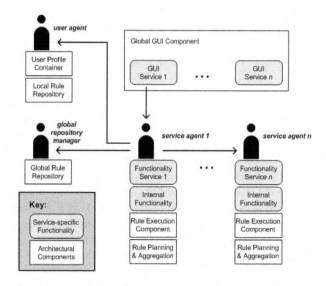

Fig. 5. Main architectural components and agents involved in an interconnection process. Arrows indicate usage of agent services.

Service and User Agents. All services available to users are based on functionality provided via agent services. The user interaction is carried out by a separate component utilizing these agent services. For purposes of service interconnection, we do not have to specify this component further. In our implemented prototype described in Section 4, it is mainly realized as an Asynchronous JavaScript and XML (AJAX)-based client-side component. Additional internal functionality (e.g. functionality required for service interconnection) is provided via agent services as well. Each main service is represented by a dedicated agent who may internally distribute requests and tasks to additional agents. Furthermore, each user is represented by a dedicated user agent, who manages and controls access to the respective user's personal profile, containing private data such as his preferences and recommended objects obtained via personalized services. It also provides a profile manager service, which may be interconnected in the same manner as any other service.

Rule Repositories. Rules are collected in and maintained by rule repositories. A global rule repository, accessible via a dedicated agent, is shared by all services involved in service interconnection processes. Currently, rules are mainly accumulated as they are provided. Ultimately, a reasoner should be able to check the consistency of the available rules, and to deduct additional rules from existing rules. In addition to the global repository, each user agent manages a local rule repository containing user-specific rules which are merged with the applicable global rules in each interconnection process.

Rule Execution Component. The rule execution component is part of all agents carrying out interconnection processes. It is utilized to apply a concrete rule to a set of input objects by resolving the rule in a straightforward manner.

Planning and Aggregation Component. The planning and aggregation component is part of all agents carrying out interconnection processes, i.e. it is utilized by the agents providing the interconnection target service. It operates on the rule repository and either attempts to determine a composition of rules, an aggregation of rules, or a combination thereof, depending on the respective interconnection process.

3 Implementation

We have implemented our approach based on Java-based Intelligent Agent Componentware (JIAC) IV [1,11], a FIPA-compliant [9,10] service-oriented MAS architecture. JIAC integrates fundamental aspects of autonomous agents regarding pro-activeness, intelligence, communication capabilities and mobility. Its scalable component-based architecture allows to exchange, add, and remove components during runtime. Standard components (which themselves can be exchanged as well) include, among others, a fact base component, an execution component, a component supporting reactive behavior, and provide individual messages to manage the appropriate actions [22]. JIAC consists of a runtime environment, a methodology, tools which support the creation of agents, and numerous extensions, such as web service connectivity, device independent interaction, and an ontology conversion engine [21]. Moreover, JIAC is the only agent-based framework that has been certified according to EAL3 of the international "Common Criteria" standard [13].

Planning and negotiation capabilities of JIAC agents are of particular interest for our approach. Invocation of agent services comprises several steps and is based on a meta-protocol, which is implemented as an extension of the FIPA request protocol. At the beginning of the service usage, an agent contacts the Directory Facilitator (DF) in order to receive a list of potential services matching the agent's current goal. If no services are found, the agent utilizes a planning component in order to determine whether the goal may be reached by combining several services partially fulfilling the goal. After receiving a list of agents providing the respective service, an optional negotiation protocol may be executed in order to determine the actual service provider. Finally, the meta-protocol covers service-independent elements such as security, accounting, communication failures, and other management-related tasks.

4   Smart Pastime Assistant

As proof of concept, we have developed and deployed the Smart Pastime Assistant (SPA)[1], a Web 2.0-based service suite integrated in a MAS. The SPA system provides various complementary agent-based services assisting users in planning their leisure time by recommending and scheduling entertainment activities. It is aimed at supporting orientation, navigation, and event scheduling in major German cities such as Berlin, Munich, and Hamburg. With the SOA approach in mind, we provide a possibility to easily add or exchange domain-specific as well as location-based information and services. Hence, it becomes possible to modify the existing prototype in order to cover alternative domains and additional or different cities and areas. The service portfolio is complemented by mapping and routing services (e.g. guided tours) and community-based services allowing users to meet other people with similar interests in real-time.

All services are personalized, i.e. they provide specific information and recommendations based on the profile of the respective user, and device-independent, i.e. they may be used on different devices as long as an AJAX-capable browser is available. An intelligent history and session management enables the user to interrupt the interaction with a service at any point and to seamlessly resume the interaction on a different device at a later time. Figure 6 gives a high-level overview of the system architecture.

In order to demonstrate our seamless service interconnection approach, we have developed an intuitive graphical user interface based on AJAX technology (including Mobile AJAX to also support mobile client devices). Here, we have specifically addressed the requirements of user-centric graphical service interconnection by utilizing universal drag-and-drop operations. Figure 7 shows a screenshot of the system with the services at a point where two services are interconnected, as described in the first scenario in the introduction, which we refer to for the evaluation of usability aspects of our approach:

Compared to a system in which services are not interconnected, the user still has to start the two services and enter search terms in order to obtain restaurants, but subsequently obtains information about the suitability of restaurants with regard to the current weather forecast after a single additional interaction, namely a drag-and-drop operation. In a system in which services are not interconnected, this information would be available only after explicitly requesting a weather forecast, and manually applying the forecast to each single restaurant result, which would require the user to carry out several additional interactions as well as to possess the required knowledge. The evaluation of our approach with regard to aspects such as development time and overall effort is more problematic, mainly because we did not implement a reference system based on hard-coded composite services. Therefore, we have restricted the evaluation to usability aspects, as described above.

---

[1] A prototype is accessible online via http://www.SmartAssistantSolutions.de.

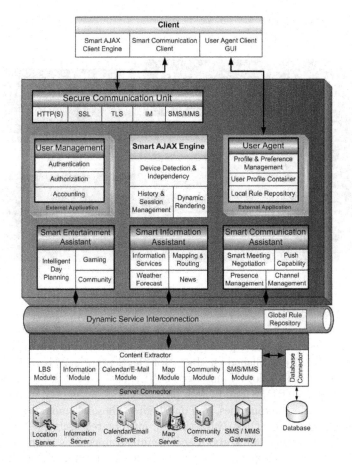

Fig. 6. High-level architecture of the Smart Pastime Assistant including the agents and components for dynamic service interconnection

5   Related Work

A large amount of work has been devoted to the area of the Service Oriented Architecture (SOA) paradigm, especially in the field of semantics-based dynamic service composition. Numerous approaches exist, ranging from automatic, dynamic, and self-organizing [3,12,16] to model driven composition of services [20]. However, actual approaches mainly focus on the composition of Web Services to new applications, such as new business processes [17]. Primarily following the paradigms of service orchestration and choreography, none of the approaches for developer-centric, consumer-centric, or user-centric service composition actually deals with dynamic user-initiated service interconnection. Most closely related are a user-centric service brokerage approach, which, however, is not based on MAS technology [24], and an approach for service composition, which allows end users to compose applications in a SOA [5].

Fig. 7. A screenshot of a service interconnection with restaurant finder service and weather forecast service

With the Resource Description Framework (RDF), the World Wide Web Consortium (W3C) established the basis for the Semantic Web with the development of the Web Ontology Language (OWL) [8]. RDF, Resource Description Framework Schema (RDFS), and OWL provide the constructs to create domain-specific vocabularies, taxonomies, and ontologies. These technologies are utilized to implement semantic SOAs. Nevertheless, current research does not address the ability to enable reasoning about service interconnections at runtime. Existing approaches focus on specifying rules in RuleML [4] and, within the context of the Semantic Web, in SWRL [14] and DRS [19]. However, the expressiveness of these languages is restricted to specifying static rules and constraints. Web service specification languages like WSDL [6] and BPEL4WS [2] provide an operational approach to service specification but do not provide any conditional relationships between source and target services in a service interconnection step. Other frameworks such as OWL-S [18] and WSMO [7], enable the specification of preconditions, postconditions, and effects. Since they are limited to static descriptions of a service's initial and final state, they are not well suited to dynamically interconnect services consisting of multiple interaction steps.

We attempt to overcome these drawbacks by combining the strengths of the above-mentioned standards with Multi-Agent technology, thus integrating adequate knowledge, negotiation, and reasoning capabilities into SOAs [15,23].

6   Conclusion and Further Work

Having identified numerous drawbacks in developer-centric service composition approaches, we have developed an agent-based architecture facilitating a user-initiated semantic interconnection of services. In this paper, we have presented several types of service interconnection as well as a rule-based approach to accomplish them. By implementing a suite of integrated entertainment services and following an AJAX-based approach for the frontend component, we have shown that it is in fact possible to semantically and dynamically interconnect individual services.

At the moment, our approach is capable of interconnecting two services per interconnection step. However, we are currently working on a generic mechanism to interconnect multiple services. For this purpose, we are enhancing the GUI to provide users with the ability to initiate a multi-service interconnection by utilizing an extended drag-and-drop mechanism. Depending on the type of service interconnection as well as on the respective source and target service, a semantically enriched context menu will be generated dynamically. This approach will allow a more flexible and powerful way of interconnecting multiple services.

Acknowledgements. We gratefully acknowledge the helpful feedback of Dr. Benjamin Hirsch, Thomas Konnerth, and Andreas Rieger, who have volunteered their time and expertise. We also thank the anonymous reviewers for their comments. Finally, we would like to thank Tarek Madany Mamlouk and Björn Stahl for contributing to the implementation of the prototype.

References

1. S. Albayrak and R. Sesseler. Serviceware framework for developing 3G mobile services. In *The Sixteenth International Symposium on Computer and Information Sciences, ICSIS XVI*, 2001.
2. T. Andrews, F. Curbera, H. Dholakia, Y. Goland, J. Klein, F. Leymann, K. Liu, D. Roller, D. Smith, S. Thatte, I. Trickovic, and S. Weerawarana. Business process execution language for web services, Version 1.1, Specification, BEA Systems, IBM Corp., Microsoft Corp., SAP AG, Siebel Systems, 2003.
3. S. Bleul and M. Zapf. Flexible automatic service brokering for SOAs. In *Proc. of the 10th IFIP/IEEE Symposium on Integrated Management*, 2007.
4. H. Boley, S. Tabet, and G. Wagner. Design rationale of RuleML: A markup language for semantic web rules. In *International Semantic Web Working Symposium (SWWS)*, 2001.
5. M. Chang, J. He, W. T. Tsai, B. Xiao, and Y. Chen. UCSOA: User-centric service-oriented architecture. In *ICEBE*, pages 248–255, 2006.
6. E. Christensen, F. Curbera, G. Meredith, and S. Weerawarana. Web service definition language (WSDL). Technical report, W3C, 2001.
7. J. De Bruijn, C. Bussler, J. Domingue, D. Fensel, M. Hepp, U. Keller, M. Kifer, B. Knig-Ries, J. Kopecky, R. Lara, H. Lausen, E. Oren, A. Polleres, D. Roman, J. Scicluna, and M. Stollberg. Web service modeling ontology (WSMO). W3C member submission, W3C, June 2005.
8. M. Dean and G. Schreiber. OWL web ontology language reference. W3C recommendation, W3C, February 2004.
9. Foundation for Intelligent Physical Agents. FIPA ACL message structure specification, FIPA00061. http://www.fipa.org/specs/fipa00061, 2002.

10. Foundation for Intelligent Physical Agents. FIPA agent management specification, FIPA00023. http://www.fipa.org/specs/fipa00023, 2002.

11. S. Fricke, K. Bsufka, J. Keiser, T. Schmidt, R. Sesseler, and S. Albayrak. Agent-based telematic services and telecom applications. *Commun. ACM*, 44(4):43–48, 2001.

12. K. Fujii and T. Suda. Semantics-based dynamic service composition. *IEEE Journal on Selected Areas in Communications*, 23(12):2361–2372, 2005.

13. T. Geissler and O. Kroll-Peters. Applying security standards to multiagent systems. In M. Barley, F. Massacci, H. Mouratidis, and P. Scerri, editors, *First International Workshop on Safety and Security in Multiagent Systems*, pages 5–16, 2004.

14. I. Horrocks, P. F. Patel-Schneider, H. Boley, S. Tabet, B. Grosof, and M. Dean. SWRL: A semantic web rule language combining OWL and RuleML. W3C member submission, W3C, 2004.

15. N. R. Jennings and M. J. Wooldridge. Applications of intelligent agents. In N. R. Jennings and M. J. Wooldridge, editors, *Agent Technology: Foundations, Applications, and Markets*, pages 3–28. Springer-Verlag: Heidelberg, Germany, 1998.

16. Y.-S. Jeon, S. Arroyo, Y.-S. Jeong, and S.-K. Han. PSM approach to web service composition. In *Proc. of WCCIA 2*, 2006.

17. F. Leymann, D. Roller, and M. Schmidt. Web services and business process management. *IBM Systems Journal on New Developments in Web Services and Ecommerce*, 41(2), 2002.

18. D. Martin, M. Burstein, J. Hobbs, O. Lassila, D. Mcdermott, S. Mcilraith, S. Narayanan, M. Paolucci, B. Parsia, T. Payne, E. Sirin, N. Srinivasan, and K. Sycara. OWL-S: Semantic markup for web services, 2004.

19. D. McDermott and D. Dou. Representing disjunction and quantifiers in RDF. In *ISWC 2002*, volume 2342 of *LNCS*, pages 250–263. Springer, 2002.

20. B. Orriëns, J. Yang, and M. P. Papazoglou. Model driven service composition. In *Proc. of ICSOC*, pages 75–90, 2003.

21. A. Rieger, R. Cissée, S. Feuerstack, J. Wohltorf, and S. Albayrak. An agent-based architecture for ubiquitous multimodal user interfaces. In *The 2005 International Conference on Active Media Technology*, 2005.

22. R. Sesseler. *Eine modulare Architektur für dienstbasierte Interaktion zwischen Agenten*. Dissertation, Technische Universität Berlin, 2002.

23. M. P. Singh and M. N. Huhns. *Service-Oriented Computing: Semantics, Processes, Agents*. John Wiley and Sons, 2005.

24. M.-R. Tazari and S. Thiergen. Servingo: A service portal on the occasion of the FIFA World Cup 2006. In *Geoinfo: Proceedings of the IWWPST 06*, pages 73–93, Vienna, Austria, 2006.

A Lightweight Agent Fabric for Service Autonomy

Yu-Fei Ma[1], Hong Xia Li[2], and Pei Sun[1]

[1] IBM China Research Lab, 19 ZGC Software Park, 8 DongBeiWangXiLu,
Haidian District, Beijing 100094, P. R. China
{mayufei,sunpei}@cn.ibm.com
[2] College of Software, Bei Hang University, 37 XueYuanLu,
Haidian District, Beijing, 100083, P. R. China
hopeshared@sse.buaa.edu.cn

Abstract. Service Oriented Architecture (SOA) is a compelling topic in Service-Oriented Computing (SOC) paradigm nowadays, as many requirements come from inter- and intra- enterprise service composition. However, as one of the most significant principles of service orientation, service autonomy, has not been addressed systematically. In this paper, we propose a feasible solution for service autonomy through analyzing its intrinsic characteristics. Firstly, from the service lifecycle management point of view, a three layer architecture of service autonomy is designed, based on which a service agent is built to provide core autonomous service functionalities, including automatic service discovery, proactive service monitoring, decentralized service orchestration, and just-in-time information sharing. Second, XMPP (eXtensible Messaging and Presence Protocol) is employed to construct a lightweight fabric of agents. Finally, three typical use cases of web service composition are used to validate the rationality and feasibility of the proposed solution for service autonomy.

Keywords: Autonomous Service Agent, Agent Fabric, Service Composition.

1 Introduction

Service orientation techniques attract more and more academic and industrial attentions in recent years, such as SOA, due to the rapid increasing of cross-organizational e-business applications. The requirements of business agility with low integration expense and high asset reusability are urgently desired in today's collaborative business environments. SOA is regarded as the key to business agility, and hailed as the glue that would bring IT closer to business. Entity aggregation embodies the business need to get a 360-degree view of those entities in one place. That implicates that autonomous services can be used to transform the way we develop systems to more closely matched business processes and solve immediate entity aggregation needs.

SOA takes the workload off the applications/services and reduce complexity in integration. So the complexity must be hidden in the messaging or communication infrastructure, say, SOA fabric. In other words, SOA pushes functionality or complexity to be handled by the messaging environment instead of the applications to be connected. In [1], *Thomas Erl* summarized eight basic principles most associated

J. Huang et al. (Eds.): SOCASE 2007, LNCS 4504, pp. 63–77, 2007.
© Springer-Verlag Berlin Heidelberg 2007

with service-orientation. Of these eight, autonomy, loose coupling, abstraction, and the need for a formal contract are considered as the core principles that form the baseline foundation for SOA. Microsoft's tenets of service orientation [2] also define four properties: service autonomy, explicit service boundaries, shared contracts and schemas, and policy-based compatibility. Therefore, service autonomy is the most representative characteristic of service orientation. However, it was not explored sufficiently and systematically compared to other principles.

The word *autonomy* has many definitions including: self-contained, self-controlling, self-governing, independent and free from external control and constraint. Service autonomy emphasizes the logics governed by a service resides within an explicit boundary. That is, the service has control within this boundary and does not depend on other services to execute its governance. It also frees the service from ties that could inhibit its deployment and evolution. Furthermore, service autonomy is a primary consideration when deciding how application logic should be divided up into services and which operations should be grouped together within a service context.

According to these assertions, the existing software services are far from autonomy, either because only simple operations are provided, such as web services, or because the services highly depend on hosting platform, such as software as a service (SaaS). A true autonomous software service can be developed, deployed, operated and managed independently, on any platform, in any organization. In order to realize service autonomy, we summarized the intrinsic characteristics from the existing literatures [1, 2, 3, 4, 5] as following:

1) Autonomous service is completely disconnected from the consumer or application. Consumers could be of various types, implemented in various technologies, and don't care about the implementation details of services. The service development, deployment, operation, management, modification and security all vary independently from those of service consumers. Autonomous service is deployed in its own execution and security environment and deployed incrementally. New services are added to a given system without breaking the functionality and services could even be deployed long time before deploying the service consumers.

2) The autonomous nature of services invariably requires processes or systems that were at one time centralized to move to a decentralized or federated model, whose individual entities communicate each other through an authorization mechanism.

3) Service autonomy also requires a greater emphasis on explicit management of trust between applications, including permission management, access control and credit mechanism. So an autonomous service needs the abilities to locate its partners and negotiate communication with them.

4) Autonomous services must also maintain control over the resource they own, because the local resource is not directly controlled by an external entity in self-governed services.

These essential characteristics are the fundamental guidelines for building a completed autonomous service ecosystem. Autonomy, in philosophy, means existence as independent moral agent, say, personal independence and the capacity to make moral decisions and act on them. In short, autonomy is a characteristic of agents. Among agents, we generally refer to social autonomy, where an agent is aware of its colleagues and is sociable, but nevertheless exercises its independence in certain circumstances, such as by refusing a request when it might harm the agent's interests. Autonomy is also in natural tension with coordination or with higher-level notions, such as commitments. To be coordinated with other agents or keep its commitments, an agent must relinquish some of its autonomy, but an agent that is sociable and responsible can still be autonomous. It would attempt to coordinate with others where appropriate and keep its commitments as much as possible, but it would exercise its autonomy in entering into those commitments in the first place.

Basically, the service-oriented applications have many similar features in the concept of software agent that is an autonomous and intelligent entity. For example, a group of agents may form a loosely coupled agent network called MAS (Multi-Agent System) [6]. They solve problems and perform tasks together through interactions and collaborations. So applying agent techniques to service orientation field is a natural choice. The research on agent-based applications has so far demonstrated that agents can glue together independently developed legacy systems. The control of a system can be distributed among autonomous agents and still maintain global coherence and capability improve greatly when systems (represented by agents) cooperate [7]. Consequently, we adopt the concepts in MAS for service autonomy architecture design in this paper. Service agent has three basic responsibilities that maintain runtime operations, manage service lifecycle and control trusty communication between agents. The first one is supported by a set of basic functions, such as service discovery, monitoring and composition. The second one is an advanced feature requiring comprehensive service modeling and governance. Agent system trustworthy is important issue for agent collaboration, which can be solved from different levels. This paper mainly focuses on the first two issues.

On the other hand, an effective communication mechanism is very important for an agent system, because autonomous systems do not stand alone without interaction with other parties. Fabric in SOA context usually means a messaging environment or communication infrastructure, which makes services or applications integrated. In this paper, we propose a lightweight agent fabric to serve the communications between autonomous service agents and, furthermore, cross-enterprise applications. According to the aforementioned autonomous system design requirements, XMPP is employed as the underlying communication and message routing technology to build this kind of lightweight fabric for agents. We also can leverage the existing XMPP technologies for the trusty communication between agents.

In general, the autonomous agent fabric can bring a number of benefits to cross-organizational collaboration, such as decoupling transactions. For traditional distributed system, solving such a scenario would be very hard. If however, all the parties are plugged into the loosely coupled agent fabric, the global transactions become local transactions and the complexity can be reduced largely. These merits would be beneficial to web services ecosystem. Web services are self-contained modular applications that can be invoked over a network, generally, the internet. As the most representative software service, it is viewed as one of the promising technologies that could help business entities to automate their operations on the web on a large scale by automatic discovery and consumption of services. The term of web service has been in place since 1999, but global interoperability still hasn't been well achieved. Though web services currently involve a single client accessing a single service, soon applications will demand federated servers with multiple clients sharing results. Such a cooperative model requires coordinative management, and it appears that an agent basis is what is needed. Agents can balance cooperation with self-interest, and they also have a property of persistence, which is necessary for establishing trust. Moreover, most agree that the web will continue to have no central authority, with its components remaining autonomous, and will support cooperative peer-to-peer interactions as well as client-server interactions. Therefore, we apply the proposed service autonomy solution to web services and validate it through service composition. The three use cases demonstrate the capabilities and effectiveness of this autonomous agent fabric, including service discovery, monitoring, decentralized orchestration and information sharing.

The remainder of the paper is organized as follows. Service autonomy architecture is introduced first in Section 2. In Section 3, a lightweight autonomous agent fabric for web service is presented in detail. The three use cases employed for validation are analyzed in Section 4. Finally, the related works are briefly summarized in Section 5, and the conclusions as well as future works are in Section 6.

2 Service Autonomy

As discussed in Section 1, we consolidate service autonomy principles into a three-level self-governance, i.e., functionality, operation and management. The self-governance, as the essential of service autonomy, demands a comprehensive service meta-model underpinning the whole lifecycle management. A completed lifecycle model provides a potent tool for eliciting the more detailed policies for governing the efficient development and usage of software services. Therefore, in this section, we designed a three-layer architecture of service autonomy based on software service lifecycle management.

From self-governance perspective, service meta-model is the information center of an autonomous agent, which maintains all management information and runtime status. The instance of service model should be created when a service is being developed and coexist with the service till it is destroyed. Here, a service-centric model is proposed to serve service autonomy by referring to the lifecycle definitions in software engineering, web services, SaaS and SOA. The whole lifecycle of software service is divided into seven phases: *demand, creation, publishing, subscribing, consuming, feedback* and *upgrading*. Among them, *upgrading* and *creation* can be regarded as one stage due to the very similar functions. In this way, the six phases constitute a cycle for continuous evolution. As shown in Fig. 1(a), the six phases of lifecycle are managed by six sub-models separately. Besides, a *service profile* sub-model provides basic description of service and contains the linkages to lifecycle sub-models. In addition, the *common service event* (*CSE*) sub-model is used to describe the events from the lifecycle which can be utilized to do performance analysis as well as problem determination among services and their interactions. *Demand* is the stage for requirement collection and analysis. Service agent can help service provider collect user requirements from service consumers or other agents. Then, the design specifications are recorded in *creation* sub-model. Also, all development information are all managed in *creation* sub-model, including bug tracing, build version, feature set, etc. When development is completed, service moves to *publishing* stage which announces the advertisements, pricing and invocation methods. When consumers subscribe a service, the corresponding subscription information and the statistic data for usage tracking is stored in *subscribing* sub-model. At *consuming* stage, the runtime and operation information are monitored. At last, the service consumers' feedbacks are collected by *feedback* sub-model, which can be used as the reference for further improvements at *upgrading* stage. In this way, all data generated in whole lifecycle are recorded in a comprehensive service model supporting service autonomy in all spectrums.

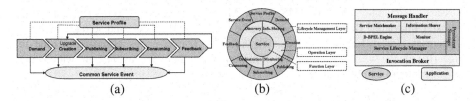

(a)   (b)   (c)

Fig. 1. (a) Service Lifecycle Management; (b) Service Autonomy; (c) Autonomous Agent

With SOA ideation, software service should not be expected to maintain those functions beyond its main business. Hence, service agent needs take the responsibilities of service operation and management in order to enable service autonomy. A three-layer agent architecture is designed to wrap simple functional services according to service lifecycle management. As shown in Fig. 1(b), the first layer is function layer provided by service itself. On top of function layer is operation layer executing operational tasks for autonomous services, including service registration/discovery, monitoring, orchestration and information sharing. The top layer is lifecycle management layer. In this manner, a software service becomes an independent entity existing on computer system or internet, managed by autonomous agent, and it can be operated individually and moved cross different platforms. Fig. 1(c) gives an abstract architecture of service agent. The basic functions of agent are provided by *Message Handler* and *Invocation Broker* modules. The former handles communication protocol parsing and message dispatch while the latter focuses on invocation adaptation and data mediation. Usually, the data from different services are converted to a uniform format, such as industry standards. The advanced operational functions are provided by other five modules: *service lifecycle manager* controlling service data; *service matchmaker* for service discovery; *monitor* collecting events and generating statistic reports; *D-BPEL engine* for service orchestration and *information sharer* providing a pub/sub mechanism for information exchange. Additionally, a persistent storage database is provided to synchronize the data with *service lifecycle manager* and *monitor* modules periodically. In this manner, service can put all operation and management workload to its autonomous agent and focus on its business logics and functionalities. Applications also can leverage agent to reduce integration complexity.

An efficient communication infrastructure is the other important aspect for service autonomy to support service composition and application integration. As both service-orientation and service autonomy requires loosely coupling, asynchronous communication is desirable for agent interactions. Asynchronous exchange allows each communication operation between two processes to be a self-contained and standalone unit of work. Each participant in a multi-step business process flow needs only be concerned with ensuring that it can send a message to the messaging system. Asynchronous interactions are a core design pattern in loosely coupled interfaces, and an asynchronous message should carry data and context around with it from place to place. Fabric, in SOA context, usually means a messaging environment or communication infrastructure, which enables service composition and application integration. Therefore, to select a suitable messaging system or communication protocol is the key to fabric design. There are many choices for asynchronous message exchange, such as JMS (Java Message Service), SMTP (Simple Mail Transfer Protocol), BEEP (Block Extensible Exchange Protocol) and so on, which are appropriate for different scenarios, respectively. The *Message Handler* module in Fig. 1(c) deals with asynchronous communication between agents, to which any proper protocol can be bound.

3   A Lightweight Agent Fabric for Web Service Autonomy

Web service is widely applied in inter- and intra- enterprise integration, however, it is still used as an interface encapsulation technique because WSDL (Web Service Description Language) only provides functional description of invocation. Based on the methodologies in Section 2, an autonomous agent can be built to enable autonomy of web service by governing the operations and lifecycle. Furthermore, XMPP is employed as asynchronous communication protocol to establish a lightweight agent fabric.

XMPP has many strong points for such a kind of autonomous web service agent system, compared to other protocols. XMPP is initially designed for XML document streaming. It defines a robust, secure, scalable, internationalization-friendly architecture for near real time messaging and structured data exchange, so that it becomes an open alternative to not only the large scale IM system, but also a wide range of XML applications. The XMPP technologies also have been extended to encompass everything from network-management systems to online gaming networks and applications for financial trading, content syndication, and remote instrument monitoring [8]. Moreover, it has been extended continuously through XEP (XMPP Extension Protocol). For example, one of important extensions is SOAP (Simple Object Access Protocol) over XMPP. Therefore, XMPP is a good choice to build a lightweight agent fabric for web services.

XMPP deals with XML document through XML streams, the containers for exchanging XML elements and stanzas. Stanzas are the first level child elements sent over those streams. Though peer-to-peer implementations exist, most of XMPP applications follow a client-server model in which a client connects and then opens a stream to the server, which causes an opening of other stream back to the client. After negotiating various stream parameters, including channel encryption and appropriate authentication, each party is free to send an unbound number of XML stanzas over the stream. To manage streams and route stanzas are XMPP server's core functions. Three primary stanza types are defined in XMPP stream and each with its own semantics. The <message/> stanza is a "push" mechanism through which one entity pushes information to another, much like email communication. The <presence/> stanza is a basic broadcast mechanism with pub/sub capabilities, through which multiple entities can receive information about a given entity to which they have subscribed. There are two elements in <presence/>; <show/> and <status/>. <show/> is used to indicate the presence of "away", "chat", "dnd" (do not disturb) or "xa" (extended away), while <status/> enables presentity post any words who would like to present. The <iq/> (stands for information query) stanza is a request-response mechanism, similar in many ways to HTTP, that allows entities to make a request and receive response from each other in an asynchronous mode. <iq/> interactions are generally used in contact-list management, configuration, feature negotiation, remote procedure calls, and other situations that require a more structured information flow than normal messaging.

In our implementation, <message/> channel is employed to exchange web service invocation message, in which the SOAP message is embedded based on a XEP specification. <presence/> channel is used as a simple pub/sub mechanism for information sharing. <iq/> channel is adopted as control channel with predefined XML controlling messages for service discovery, event retrieval, D-BPEL deployment and information sharing. We leverage open source projects, *Wildfire* [9] and *Smack APIs* [10], to build a coordination server and an autonomous agent of this fabric, respectively. Fig. 2(a) is the implemented autonomous agent for web service, in which SOAP codec is enhanced to support WS-Addressing specification [11], so that SOAP message can be delivered to the third part, instead of always between two agents. Similarly, discovery and Pub/Sub modules are all implemented following XEP specifications. Besides, we also provide a tool for service preview, called service browser (Fig. 2 (b)), which is like a client of instant messenger.

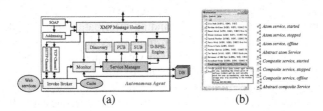

(a)   (b)

Fig. 2. (a) Autonomous Agent for Web Services; (b) Service Browser

3.1 Service Discovery

Although agents communicate is like a peer-to-peer manner, the agent fabric is client-server architecture. The enhanced *Wildfire* server plays a role of coordinator and the agents are its clients. Each agent needs register on a coordinator to obtain a unique ID. The coordinators take charge of the communication with agents and other coordinators in the fabric. During the service discovery in a decentralized system, the request usually needs a broadcast mode to probe remote ends. In order to reduce the communication workload in fabric, a service pre-filtering process is conducted in coordinators leveraging registration information. So coordinators can filter out most of irrelevant services and only forward the discovery request to a small proportion of candidate agents where the precise matching is executed based on the detailed descriptions in service model.

As shown in Fig. 3(a), when a service needs look for a partner in fabric, its agent sends a discovery request to its local coordinator where it registered. The local coordinator, on one hand, forwards the request to other coordinators for cross domain discovery. On the other hand, at local coordinator, the registered services are pre-filtered according to category and keywords. Then, the coordinator only forwards the request to the filtered services for advanced matching. The remote coordinators also deal with the request in the same way. When a candidate service receives the request, it compares the items in request message to the corresponding items in its own service model, including detailed description, finer categorization, semantic reasoning, service interfaces and even SLA (Service Level Agreement) negotiation, etc. If the service is matched, a successful matching notification needs to be sent back to the requester, so that the service requester can obtain its ID and invoke this service.

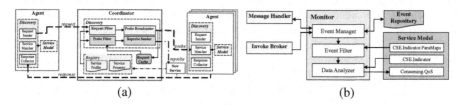

(a)   (b)

Fig. 3. (a) Service Discovery Module; (b) Service Monitor Module

Here are the sample massages of request and response during service discovery. In request message, there are two parts of properties. One is the basic description of service, which is used by server to filter out candidate service at a coarse level, including name, version and category. The rest part includes the descriptions of input/output, functions, precondition and binding information, etc. Besides, query id and expire date are two import elements in request. A unique ID is required in the corresponding response message so that service requester can differentiate the received responses from different conversations. Expire date is used to indicate when the request should be deleted from the request repository on coordinator. In contrast, the response is a simple message to notify the result to the requester, which contains query ID, service name and service ID and an optional brief description of service.

```
Service discovery request:
<iq type="get" from="weatherod@example.com" to="discovery.example.com">
  <discovery xmlns="crl:ibm:service:discovery">
    <name>Weather</name><version>1.0</version>
    <categories><category>Weather</cateory>...</categories>
    <interface>
      <inputDesc>...</inputDesc><outDesc>... </outDesc>
      <functionDesc>Weather Report</functionDesc>
      <preconditionDesc>...</preconditionDesc><wsdl>...</wsdl>
      <localname>...</localname><porttype>...</porttype>
    </interface>
    <queryid>12345</queryid><expireDate>2006-11-9 13:00:00</expireDate>
  </discovery>
</iq>
```

```
Service discovery response:
<iq type="result" from="weathernetwork@example.com"
            to="weatherod@example.com">
  <discovery xmlns="crl:ibm:service:discovery">
    <name>The Weather Network</name>
    <queryid>12345</queryid>
  </discovery>
</iq>
```

In addition, a *reprobe* process is used to automatically discover those newly registered services referring to WS-Discovery specification [11]. That is, coordinator maintains a request repository, where the unexpired service requests are maintained. When a new service is registered on this server, the requests in the repository will be filtered to verify if this new service meets the requirements of those already existing requests. If there is any, coordinator will *reprobe* this new service using these candidate requests for advanced matching. Meanwhile, if the once matched services are not available any more, the service discovery request will be sent out again to seek new ones automatically.

3.2 Service Monitoring

Monitoring is the core of service lifecycle management, as it collects all events generated by the activities in all functional modules. It also provides data analysis and reporting functions. Fig. 3(b) gives the details of monitor module which is composed of three sub-modules: event manager, event filter and data analyzer. The events from different modules need to be stored in database through event manager, which also processes the event retrieval requests from other agents. On the other hand, event manager passes each event received to event filter where the useful data for analysis are extracted according to the parameter mapping policies. Data analyzer is a computational module for statistical report generation. It can integrate different data analysis tools for reporting. In service meta-model, four QoS parameters are defined: *average response time, failure rate, mean time to repair* and *mean time between failures*. The computation formulas are specified by the *indicators* of *CSE*. There are six types of *CSE* corresponding to the lifecycle phases. A number of actions are defined in each type of *CSE*. For example, *demand* actions include *open demand, close demand, change demand, get demand*, etc. The average response time and failure rate of service can be calculated based on invocation events, which contains the properties of *start time, end time, state of failure, requester id* and *response receiver id*. In addition, agents can retrieve the events stored in remote agents if they have trustworthy relationship. Following is a sample process of retrieving an invocation event.

Event retrieval request:
```
<iq type="get" from="tripod@example.com"
          to="yahootravel@example.com">
  <eventquery xmlns="crl:ibm:service:eventquery">
    <trigger>weathernetwork@example.com</trigger>
    <type>CONSUMPTION</type>
    <action>invoke</action>
    <starttime>2006-11-9 11:00:00</starttime>
    <endtime>2006-11-9 13:00:00</endtime>
  </eventquery>
</iq>
```

Event retrieval response:
```
<iq type="result" from="yahootravel@example.com" to="tripod@example.com">
  <eventlist xmlns="crl:ibm:service:eventlist">
    <event xmlns=" crl:ibm:service:event">
      <trigger>weathernetwork@example.com</trigger>
      <type>CONSUMPTION</type>
      <action>invoke</action>
      <timestamp>2006-11-9 12:00:00</timestamp>
      <x xmlns="jabber:x:data'" type="form">
        <title>Invocation Event</title>...
        <field type="text-single" var="startTime">
          <value>2006-11-9 12:00:00</value>
        </field>...
      </x>
    </event>...
  </eventlist>
</iq>
```

3.3 Decentralized Orchestration

A traditional BPEL program invokes services distributed on different servers; however the orchestration of these services is typically under centralized control, which is opposite to the philosophy of service autonomy. Moreover, performance and throughput are the other major concerns. Therefore it is necessary to decentralize control of BPEL engine. In this work, we implemented a decentralized BPEL execution mechanism using autonomous agents by employing the BPEL partition technique in [12] and the decentralized BPEL engine (D-BPEL). During the BPEL editing, the composite service is specified by traditional centralized BPEL program, first. Then this program is partitioned into independent sub-programs that can interact with each other without any centralized control. These sub-programs need to be installed on the D-BPEL engines of corresponding agents. That is, all web services involved in a composite service only need execute a small segment of BPEL program within its own context without need for knowing the overall relationship of orchestration. Particularly, the dynamic service binding is controlled by the master service within a composition context, where the D-BPEL deployment is managed. If the participant services change, the master service needs update or re-deploy the D-BPEL sub-programs to new ones. Besides supporting service autonomy, decentralized orchestration can increase parallelism and reduce network traffic required for an application, as well as minimize communication costs and maximize the throughput of multiple concurrent instances of the input program. <iq/> channel is utilized to deploy D-BPEL programs to the participant services as follows:

```
D-BPEL program deployment request:
<iq type="set" from="tripod@example.com"
          to="weathernetwork@example.com">
  <deploy xmlns="crl:ibm:service:deploy">
    <serviceid>tripod</serviceid>
    <servicename>Trip On Demand<servicename>
    <dbpel>...<dbpel>
  </deploy>
</iq>
```

```
D-BPEL program deployment confirmation response:
<iq type="result" from="weathernetwork@example.com"
          to="tripod@example.com">
  <deploy xmlns="crl:ibm:service:deploy">
    <serviceid>tripod</serviceid><status>success</status>
  </deploy>
</iq>
```

4   Use Cases for Validation

Three use cases are simulated to illustrate the usage of the autonomous agent fabric for web services and validate its rationality, feasibility and benefits. Fig. 4(a) shows a fabric architecture used for use cases simulation and validation.

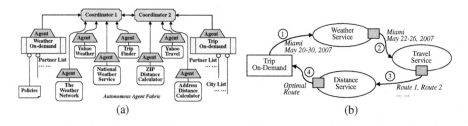

(a)   (b)

Fig. 4. (a) Fabric Architecture for Use Case Validation; (b) Decentralized Orchestration

Use Case 1: Mike develops a desktop application to show the weather condition in real time by retrieving weather report from web services, called *Weather On-demand*. In order to guarantee weather information is always available, Mike plans to maintain a web service list in the application and select the best service to invoke when the weather report needs to be updated. With autonomous agent fabric, to realize this kind of application becomes much easier than before. Firstly, Mike needs develop an

Eclipse RCP application using the Agent SDK we provided. Then, register this application in the fabric via agent APIs. The weather service selection criterion can be configured as a set of policies in application, according to which a service discovery request will be automatically generated and sent out by agent. Supposing some weather reporting services already connected to the fabric through agents, including The Weather Network [30] and National Weather Service [31] matching the discovery request. So the two services send back successful matching notification to the application where they are added into the partner list and their presence information is subscribed. If any service gets offline or busy, the other service will be selected to invoke. Later, Yahoo Weather [32], as a new service, is connected to the fabric. It will automatically receive the weather service discovery request cached in Coordinator's request repository. If Yahoo Weather matches the request, it also becomes a candidate service in the partner list of application. The automatic service discovery in agent fabric is helpful for service availability assurance. In this manner, *Weather On-demand* can always obtain new services meeting requirements and select the best service to invoke according to the presence status and QoS indicators.

Use Case 2: Mike extends his application to support travel arrangement according to weather condition, say, provide travel route recommendations within the continuous days of good weather, called *Trip On-demand*. The input will be the destination and a range of available days; and the output will be a list of possible travel routes with specific time arrangement. Supposing the two travel services, Yahoo Travel [33] and Trip Finder [34], are connected to the agent fabric, and *Trip On-demand* has found them by auto-discovery like Use Case 1 shows. This time, the application has five service partners, three for weather report and two for travel arrangement. Mike input Hawaii as destination parameter to test the service. However, he does not get any response. In order to debug, Mike uses agent APIs to retrieve the history events from event repository in the invoked services to determine the problem. It was found that The Weather Network is a local weather service for Canada and does not provide Hawaii's weather report. So Mike adds a new service selection policy to application configuration that if there is an error or timeout notification then the other services in the list are automatically tried sequentially until a success response is obtained. Another choice to solve this problem is to add service selection logics to the application according to the region coverage in each weather service. Here, we suppose all weather report services publish the covered city list via Pub/Sub mechanism. In this way, the application can subscribe the city list of weather services. If any city is added to or removed from the list, the application will receive a notification timely so that it can dynamically select the best service to invoke. This use case demonstrates the benefits from service monitoring as well as just-in-time information sharing.

Use Case 3: Mike further extends *Trip On-demand* by adding attraction arrangement, so that it can provide recommendations of attraction ordering according to the minimum travel distances. Here, a distance calculation service is added. A number of distance calculation services are available on Internet, such as StrikeIron Address Distance Calculator [35] and ZIP Distance Calculator [36]. The zip codes of two locations are the input parameters of the distance services while the distance between them in mile is the returned result. Again, we suppose the two distance services already enroll in agent fabric. In this case, a decentralized orchestration is adopted to enhance service autonomy and reduce service composition complexity. That is, the composition logic is segmented into three D-BPEL programs which are deployed on the agents of weather, travel and distance services, separately. In this decentralized manner, *Trip On-demand* only collects end user's input parameters and showing the final result of travel routes, instead of receiving all mediate results and control all business logics. As shown in Fig. 4(b), the application sends user request to a weather service to select a period of time with good weather condition at travel destination first. Then, travel service is directly called by weather service to select appropriate travel routs according to the time constrains. Sequentially, the recommended travel routes are passed to distance service to calculate distances between any two attractions. Finally, the travel route with the optimal attraction order is returned to the application. In whole process, no central controlling BPEL is needed and the redundant data transmissions are moved.

In the three use cases, we simulated the composition of the real web services on the Internet. All service invocations are brokered by autonomous agents, whether they have SOAP interfaces, REST APIs or just HTTP requests. The experiments show the proposed fabric is effective and easy-to-use for web service autonomy and composition. Meanwhile, this fabric also provides a semi-open ecosystem for cross-organization collaborations, which is well governed and can be easily scaled out.

5   Related Works

Service oriented computing (SOC) is an emerging cross-disciplinary paradigm for distributed computing that is changing the way software applications are designed, architected, delivered and consumed. The basic disciplines of SOC were addressed in many literatures. In [1], *Thomas Erl* gave a good summarization of service orientation. Service Oriented Architecture (SOA) is a form of distributed system architecture, and its six typical properties are consolidated by W3C working group of web service architecture [13]. Services are autonomous and platform-independent computational elements that can be used to build networks of collaborating applications distributed within and across organizational boundaries. An agent is an encapsulated computer system that is situated in some environment and that is capable of flexible, autonomous action in that environment in order to meet its design objectives. A software agent is a piece of software that autonomously acts to perform tasks on user's behalf [6]. Agents are often viewed as the synonymous of services in principle. However, agent is a broader concept. In [14], the notion of agent-based web services (AWS) is proposed, including architecture and meta-model and integration. The key challenge is to develop an integration framework for the two paradigms, agent- and service-oriented, in a way that capitalizes on their individual strengths. There are a series of workshops (WSABE/SOCABE) are bringing SOC (or web services) and agent areas together in recent years [15, 16, 17]. The point of convergence between these two fields allows us to choose the paradigm of cooperative agents to model e-business applications.

The faculty of agent enables enterprise set up efficient communication mechanism and relationship with their customers, their suppliers and their partners. Multi-Agent

System (MAS) has been applied to e-Business field for application coordination and control in an autonomous way. In [15], the agents are used as proxies that assist web services selection according to some criteria. The web services composition also can leverage agent system [16]. The CB-SeC framework [17] is another sample of agent-based architecture that provides service selection and composition in pervasive computing environment. In [18], context-aware agents are introduced to sense the context (such as computing resources, user type and physical environment) and optimize service composition. On the other hand, SOC applications require traditional MAS more open for cooperation purpose. In [19], a cooperative multi-agent architecture is proposed, based on which a better solution is formed by integrating web services [20]. Furthermore, a multi-agent based framework is built to facilitate testing web services in a coordinated and distributed environment in [21].

Service modeling is an inevitable problem to build SOA. A variety of service models have been defined from different perspectives, such as the ones in [22] and [23]. The report [22] urged that a service needs to be managed throughout its lifecycle, and that service management should not be a consideration that is limited to deployed services. Web services are self-contained and self-describing modular applications that can be published, located, and invoked across the web. Web Services are the current most promising technology for SOA. WSDL is a basic meta-model for web service description, which is the key to making web service applications loosely coupled. WSDL defines the interfaces and methods necessary for accessing the web services. UDDI (Universal Description, Discovery and Integration) [11] is also a meta model as an industry standard that allows businesses to describe and register their web services, as well as discover the services that fit their requirements. In recent years, semantic web technologies are employed to enhance the description capabilities of web services, such as OWL-S (Web Ontology Language for Services) [24]. The OWL-S is the ontology of services and designed for the communications among applications that need to process the semantic information content. To combine different description models is a trend for future web services. The literature [25] and [26] give a good reference for the integration of OWL-S with UDDI, and WSDL with UDDI, respectively, while the relationship between WSDL and OWL-S is described in [24]. Due to the high complexity in SOA, service monitoring is a challenging issue. One of fundamental problems is QoS tracing, which has been investigated by many research works, especially in workflow area. For example, In [27], a QoS oriented framework: WebQ was presented, which is composed of several QoS parameters, such as latency, throughput, reliability, availability, cost, etc. On the other hand, information systems are becoming increasingly autonomous, so that the service composition begins to move to the decentralized from central controlling. *Benatallah et al.* describe a peer to peer execution pattern for orchestrating web services to overcome the bottleneck associated with having a centralized controller [28]. In this work, we realized a decentralized BPEL execution mechanism in agent by employing the BPEL partition technique in [12].

To sum up, the problems involved in this paper, such as, the service autonomy, agent-based architecture, service composition, etc. are addressed more or less by other research work. However, systematical solution of service autonomy is not investigated sufficiently. *Colombo* project [29] tried to build a comprehensive platform for service-oriented applications to provide deep exploratory of how middleware can address the requirements of the SOC paradigm. However, *Colombo*'s approach is to build a native web services runtime and programming model focused on the execution and development of service-oriented applications. In contrast, our work focuses on the most distinctive feature of SOC: autonomy, and seeks for a feasible solution by combining cross-disciplinary methodologies, including SOA, web services, agent system and instant messaging technologies.

6 Conclusions

In this paper, we present an end-to-end solution for service autonomy. Based on service lifecycle management, a lightweight autonomous agent fabric is built over XMPP messaging architecture. We apply this framework to web services, and justify its values by three user cases. The rationality and feasibility of the system has been validated through the simulation results. It is an interesting attempt to combine the strengths of SOA, web services, agent system and instant messaging technologies. We believe that the keys to the next-generation web are cooperative services, system trust, and semantic understanding, coupled with a declarative agent-based infrastructure. Furthermore, the proposed agent fabric could be more concise if we abstract a middleware with common functions from software engineering perspectives. However, it is beyond the scope of this paper.

References

1. T. Erl. Service-oriented Architecture: Concepts, Technology, and Design. Prentice Hall PTR, NY, US. Aug. 2, 2005.
2. M. Burner, Service Orientation and Its Role in Your Connected Systems Strategy, MSDN, July 2004.
3. L. F. Cabrera, C. Kurt, D. Box, An Introduction to the Web Services Architecture and Its Specifications, version 2.0, MSDN, Oct. 2004.
4. U. Dahan, Autonomous Services and Enterprise Entity Aggregation, MSDN, July 2006.
5. D. Ferguson, T. Storey, B. Lovering, J. Shewchuk, Secure, Reliable, Transacted Web Services, IBM developerworks, Oct 2003.
6. N. R. Jennings. On Agent-based Software Engineering. Artificial Intelligence, 117(2): 277-296, 2000.
7. M.N. Huhns, D.A. Buell, Trusted Autonomy, IEEE Internet Computing, Volume 6, Issue 3, May-June 2002 Page(s):92 – 95.
8. P. Saint-Andre, Streaming XML with Jabber/XMPP, IEEE Internet Computing, Volume 9, Issue 5, Sept.-Oct. 2005, Page(s):82 – 89.
9. Wildfire: http://www.jivesoftware.org/wildfire/
10. Smack API: http://www.jivesoftware.org/smack/
11. http://en.wikipedia.org/wiki/List_of_Web_service_specifications.
12. M. G. Nanda, S. Chandra, and V. Sarkar, Decentralizing Execution of Composite Web Services, in Proc. of OOPSLA, Vancouver, B.C., Canada, Oct. 24-28, 2004.
13. Web service Architecture: http://www.w3.org/TR/2004/NOTE-ws-arch-20040211/. W3C Working Group Note, February 2004.
14. Y. S. Li, H. Ghenniwa, et. al. Agent-Based We Services Framework and Development Environment. Computational Intelligence, Vol. 20, No. 4, 2004.

15. E. M. Maximilien and M. P. Singh. Agent-based Architecture for Automatic Web Service Selection. in Proc. of the 1st International Workshop on Web Services and Agent based Engineering, July 2003.
16. D. Richards, et. al. Composing Web Services using an Agent Factory. in Proc. of the 1st International Workshop on Web Services and Agent based Engineering, July 2003.
17. S. K. Most´efaoui and G. K. Most´efaoui. Towards A Contextualisation of Service Discovery and Composition for Pervasive Environments. in Proc. of the Workshop on Web-services and Agent-based Engineering, July 2003.
18. Z. Maamar, S. K. Mostefaoui, and H.Yahyaoui, Towards an Agent-based and Context-oriented Approach for Web Services Composition. IEEE Trans. on Knowledge and Data Engineering, Vol. 17, No. 5, 686-697, May 2005.
19. M. Brahimi, et. al. A federated Agent based Solution for Developing Cooperative E-Business Applications. in Proc. of the 4th international workshop in Web based Collaboration, Zaragoza, Spain, 2004.
20. M. Brahimi, M. Boufaida, L. Seinturier, Integrating Web Services within Cooperative Multi Agent Architecture. in Proc. of the Advanced International Conference on Telecommunications and International Conference on Internet and Web Applications and Services, 2006.
21. X. Y. Bai, et. al. A Multi-Agent based Framework for Collaborative Testing on Web Services. In Proc. of the 4th IEEE Workshop on Software Technologies for Future Embedded and Ubiquitous System and 2nd International Workshop on Collaborative Computing, Integration, and Assurance, 2006.
22. John Dodd. The Service Lifecycle. CBDI Journal, Nov. 17, 2005.
23. H.T. Xia, L. M. Meng, and X. S. Qiu, A Generic Lifecycle-based Service Management Information Modeling. in Proc. of the Eighth IEEE International Symposium on Computers and Communication, 2003.
24. D. Martin, M. Burstein, et al. OWL-S: Semantic Markup for Web Services. http://www.w3.org/Submission/OWL-S/
25. D. Martin, et al. Bringing Semantics to Web Services: The OWL-S Approach. Semantic Web Services and Web Process Composition. First International Workshop, 2004.
26. A. T. Manes. Registering a web service in UDDI. SOA Web Services Journal, Vol. 3, Issues 10. 6–10, 2003.
27. C. Patel, K. Supekar and Y. Lee. A QoS Oriented Framework for Adaptive Management of Web Service Based Workflows. Lecture Notes in Computer Science, (Springer-Berlin Heidelberg 2003), 2736: 826–835.
28. B. Benatallah, M. Dumas, et. al. Overview of Some Patterns for Architecting and Managing Composite Web Services. In ACM SIGecam Exchanges, volume 3.3, pages 9-16, 2002.
29. F. Curbera, M. J. Duftler, R. Khalaf, et al. Colombo: Lightweight Middleware for Service-Oriented Computing. IBM System Journal, Vol. 44, No. 4, 2005.
30. http://theweathernetwork.com/weather/index.htm
31. http://www.weather.gov/
32. http://weather.yahoo.com/
33. http://travel.yahoo.com/
34. http://www.iexplore.com/trip/trip.jhtml
35. http://www.strikeiron.com/ProductDetail.aspx?p=136
36. http://webservices.imacination.com/distance/index.jsp.

Semantic Service Composition in Service-Oriented Multiagent Systems: A Filtering Approach*

Alberto Fernández and Sascha Ossowski

Universidad Rey Juan Carlos. Tulipán s/n, 28933. Móstoles (Madrid) - Spain
{alberto.fernandez,sascha.ossowski}@urjc.es

Abstract. In Service-Oriented MAS middle agents provide different kinds of matchmaking functionalities. If no adequate services are available for a specific request, a planning functionality can be used to build up *composite* services. In order to take advantage of recent advances in the field of AI planning for this purpose, we propose exploiting organisational information of Service-Oriented MAS to heuristically filter out those services that are probably irrelevant to the planning process. We present a novel framework for service-class based filtering and show how it can be instantiated to a particular MAS domain based on role and interaction ontologies.

1 Introduction

Services are computational entities that can be described, published, discovered, orchestrated and invoked by other software entities. When the management of services is realised by agents, the term service-oriented MAS has become popular [11,12].

Service-oriented architectures usually include directory services that service providers register with. A request for a service with desired characteristics often results in a matchmaking process based on the profiles stored in the directory. In service-oriented multiagent systems, this functionality is typically offered by middle-agents [6].

Even if no adequate services are found in this manner, it is still possible to build up *composite* services by purposefully combining several pre-existing services. In order to cope with this challenge, a middle-agent needs to be endowed with a planning capability that receives service profiles as input and orchestrates them in a desired manner. The service composition planning problem has subtle differences with the classical AI planning problems, in particular as service composition plans need not be very deep but, in turn, can be built up from a vast number of services (operators) that are usually registered in the directory.

---

* This work has been partially supported in part by the European Commission under grant FP6-IST-511632 (CASCOM), and by the Spanish Ministry of Education and Science, projects TIC2003-08763-C02-02 and TIN2006-14630-C03-02

This is particularly the case in open large-scale service-oriented MAS, where a pure AI planning [14] approach can become impracticable.

In order to still take advantage of the recent advances in the field of AI planning, we suggest to reduce the set of input services that are passed on to an agent's composition planning component. In particular, in this paper we show how to make use of the organisational structure underlying a MAS in order to heuristically filter out those services that are probably irrelevant to the planning process.

The paper is organised as follows. We first describe a generic framework for service-class based filtering. We then show how in service-oriented MAS this service class information may be obtained from role and interaction taxonomies that are derived from the MAS organisational model. Our approach is illustrated in the emergency healthcare domain of the European IST project CASCOM [2,9]. Finally, we present the lessons learnt from this enterprise and point to future lines of work.

2   Generic Filtering Framework

In this section we first give a birds-eye view of our approach to service filtering for composition: setting out from an ideal situation, we outline the major difficulties that need to be overcome, so as to motivate our service filtering mechanism. The details of the process are described in subsequent subsections.

2.1   Overview

At a high level of abstraction, the service composition planning problem can be conceived as follows: let $P = \{p_1, p_2, ..., p_m\}$ be the set of all possible plans [1] for a given service request R, and $D = \{s_1, s_2, ..., s_n\}$ the set of input services for the proper service composition planner (i.e. the directory available). The objective of a filter F is to select a given number l of services from D, such that the search space is reduced, but the best plan of P can still be found.

In an *ideal situation* with complete information (i) the set of all plans P is known and (ii) the *quality* of each plan can be evaluated (obviously, the plan that requires the least number of services is not always the one that best matches a query). In this case, the set of services returned by the filter should include all the services of the best plan (as well as some others until the number of l services is reached). However, it is obvious that this ideal case is not realistic since the problem would be already solved (i.e. the plan of maximum quality is supposed to be known beforehand).

In a next step, suppose that it is not possible for the filter to evaluate the quality of the plans in P. In this case, if the number of services necessary for the execution of all plans in P is bigger than the number of services l that are

---

[1] We consider a *plan* as a composition of two or more services connected by some control constructs such as sequence, if-then-else, and so on. In particular, we rely on the OWL-S service model.

allowed to pass our filter, the latter should make sure that the pruning of the search space for the planner is minimal. Put in another way: the bigger the subset of plans $P' \subset P$ that the planner can choose from, the bigger the probability that the plan of maximum quality is among them. Therefore, the filter should select those services that maximise the cardinality of P', i.e. that maximise the number of plans from P that are available to the planner. A good heuristic to this respect is based on *plan dimension* and on the *number of occurrences* of services in plans: a service is supposed to be the more important, the bigger the number of plans from P that it is necessary for, and the shorter the plans from P that it is required for.

Again, it is unrealistic to assume complete information respecting the set of plans P for a given query R in general, and respecting their length and number of occurrences of services in them in particular. Nevertheless, we can approximate this information by storing and processing previously created plans. So, in principle, we can build up matrices as the shown in Table 1 for every possible query. In the example, for some query R, service s_2 was part of 10 plans composed by two services and 55 plans formed by 3 services. In the example, the matrix also stores the number of plans generated of each dimension.

Table 1. Example of information about historical plans

Historical information about plans for a service request R				
Dimension	1	2	3	...
# of plans	0	50	70	
s_1	0	7	24	
s_2	0	10	55	
s_3	0	13	10	
s_4	0	17	11	
...				

However, it soon becomes apparent that the number of services and possible queries is too big to build up all matrices of the above type. The memory requirements would be prohibitively high and the filtering process would become computationally too expensive. Furthermore, the continuous repetition of a very same service request R is rather unlikely. And, even more important, this approach would not be appropriate when a new service request (not planned before) is required (which, in fact, is quite usual).

To overcome this drawback, we assume the availability of *service class information*, so as to cluster services based on certain properties. So, there will only be matrices for each *class* of service request (query), and the matrix information stored for *classes* of services instead of services. If the number of classes is not too big, the aforementioned approach becomes feasible computationally.

Service classes need not be disjoint. In particular, for service advertisements we may specify that a service belongs simultaneously to several different classes, and for service requests we allow a formula in disjunctive normal form (i.e. a disjunction of conjunction of classes)[7].

Figure 1 depicts the structure of our approach to service composition filtering. With each outcome of a service composition request, a Historical Information Matrix H (an abstraction of Table 1) is updated. Setting out from this information, a Relevance Matrix v is revised and refined. Based on this matrix, service relevance can be determined in a straightforward manner. For each service composition request, the filtering method is based on this estimated service relevance function.

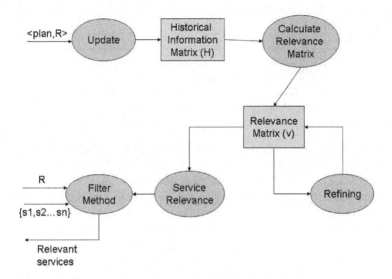

Fig. 1. Architecture of the filter component

2.2 Obtaining the Relevance Matrix

We now describe how the relevance matrix $v(s, r)$ can be obtained from past plans. We first describe how the information about plans is stored and how the relevance matrix is calculated from this information. Then, we propose a method to refine the relevance matrix. Finally, several options for bootstrapping are described.

Historical Information About Plans. The *relevance matrix* v represents the estimation of a *service class* s to be included in a (the best) composite plan to provide the requested *service class* r. In order to create the relevance matrix we set out from a set of plans (composite services) that were created in the past. We deal with simplified versions of those plans. In particular we are only interested in the *classes of services* that compose every plan. More precisely, for each plan we use the following information:

a) Request classes expression (disjunctive normal form) included in the request. We denote C_{R1}, C_{R2}, ..., C_{Rn} the classes included in that expression.
b) Plan dimension and classes: the set of classes $PC = \{C_{P1}, C_{P2}, ..., C_{Pm}\}$ obtained after mapping the services included in the composite plan.

The information about past plans created is stored in a matrix similar to Table 2 (adapted from Table 1 by considering classes instead of services, and where, for instance, services s_3 and s_4 have been clustered in class C_3). Matrixes H^R of this type exist for every service class R.

Table 2. Example of class information about historical plans

H^R: Historical information about plans for service class R (Request)				
Dimension	1	2	3	...
# of plans	0	50	70	
C_1	0	7	24	
C_2	0	10	55	
C_3	0	30	21	
...				

The information in this table is computed as follows. If only one service class is included in the request expression the value of the "# of plans" and every class of PC is incremented by one, for that plan dimension[2].

However, when more than one class are present in the request expression, things are not that simple. The problem is that, when there is more than one class in the request, it is not straightforward to determine the parts of the request that the service (classes) of the composite plan are relevant for.

In the case of a conjunction expression (e.g. $C_{R1} \wedge C_{R2}$), we assume that all the classes in the plan are relevant for both C_{R1} and C_{R2}, since both conditions must be fulfilled. However, in the case of a disjunction this is not the case. For instance, if the request expression is $(C_{R1} \vee C_{R2})$ and the plan includes the classes $\{C_{P3}, C_{P4}, C_{P5}\}$, it is not clear whether these three classes are participating in providing both C_{R1} and C_{R2}, or if, for instance, C_{P3} and C_{P4} are relevant for C_{R1}, and C_{P5} only for C_{R2}. In this last case we decrease the possible negative impact of adding one unit to the wrong class by weighting the contribution of the plan information by the inverse of the number of terms in the disjunction expression. For example, suppose a query (request expression) of the following shape: $(C_{R1} \vee (C_{R2} \wedge C_{R3}) \vee C_{R4})$. Furthermore, assume that the classes of services required by the plan (composite service) that have been determined to best match this query are $PC = \{C_{P1}, C_{P2}\}$. According to our model, the dimension of that plan is 2, and the historical matrixes corresponding to the four classes in the request expression (C_{R1}, C_{R2}, C_{R3}, C_{R4}) are updated in line with this. In particular, the values of $H^{C_{R1}}($"# of plans",2), $H^{C_{R1}}(C_{P1}, 2)$, $H^{C_{R1}}(C_{P2}, 2)$, $H^{C_{R2}}($"# of plans",2), $H^{C_{R2}}(C_{P1}, 2)$, $H^{C_{R2}}(C_{P2}, 2)$, and so on for $H^{C_{R3}}$ and $H^{C_{R4}}$, are incremented by $\frac{1}{3}$ as the number of disjunctive terms in the query is 3. Figure 2 shows the algorithm for updating the historical matrix.

---

[2] As all services need to be available for the plan to be executable, every service class has the same relevance for the plan, independently of whether one or several services of that class are used by the plan.

```
<PC> = SET OF Class          // plan classes
<RE> = <DisjunctionExpr>     // request expression
<DisjunctionExpr> = SET OF <ConjunctionExpr>
<ConjunctionExpr> = SET OF Class

UpdateHistorical(H: historical matrix;RE: request expression;PC: plan classes)
{
  dim = Card(RE)
  FOR ALL Conj IN RE {
    FOR ALL r IN Conj {
      H^r("# of plans",dim) = H^r("# of plans",dim) + 1/dim
      FOR ALL p IN PC {
        H^r(p,dim) = H^r(p,dim) + 1/dim
      }
    }
  }
  return H
}
```

Fig. 2. Historical information matrix update algorithm

Calculus of the Relevance Matrix. As commented above, by ranking services we try to select a set of services that covers the largest subset of the plan space, as an attempt to maximise the chance of the best plan to be contained in it. Services that formed smaller plans in the past are considered more relevant, since it is easier to cover small plans than large ones, so with less services more plans can be covered.

We use the following function to aggregate the information about plans (remember that all this information is about a single request class R):

$$Relevance(C, R) = \frac{\sum\limits_{d=1}^{m} \frac{n_d}{d^c}}{\sum\limits_{d=1}^{m} \frac{N_d}{d^c}} \tag{1}$$

where d is the dimension of the plan, m is the dimension of the longest plan stored, n_d is the number of times that C was part of a composite plan of dimension d for the request R, and N_d is the total number of plans of dimension d ("# of plans") for that request. Note that each appearance of class C in a plan contributes to the relevance value, and that this contribution is the higher the smaller the plan dimension. c is a constant > 0 that allows adjusting the level of importance of plan dimensions.

In the example of Table 2,

$$Relevance(C_1, R) = \frac{7/2 + 24/3}{50/2 + 70/3} = 0.24 \tag{2}$$

With this calculus we obtain a relevance value between 0 and 1 for every given service class C with respect to the composition of a service of class R.

Refining the Relevance Matrix. The matrix $v(s,r)$ specifies the relevance of a service class s to be part of a plan (composite service) that matches the query for a certain service class r. However a situation like the following may occur: Suppose that a plan that achieves C_1 is searched for, and that a potential solution is to compose the services C_2 and C_3 ($C_2 \oplus C_3$ for short). However there is no service provider for C_3, but instead C_3 can be composed as $C_4 \oplus C_5 \oplus C_6$, so the final plan is $C_2 \oplus C_4 \oplus C_5 \oplus C_6$. Unfortunately, the value $v(C_4, C_1)$ is low and the service providing C_4 is discarded and not taken into account in the planning process, so the aforementioned plan cannot be found by the planner. Therefore, we will refine the relevance matrix by taking *transitivity* into account, e.g. through the following update: $v(C_4, C_1) = v(C_4, C_3) \cdot v(C_3, C_1)$. The same holds for third-level dependencies (e.g.: $v(C_7, C_1) = v(C_7, C_4) \cdot v(C_4, C_3) \cdot v(C_3, C_1)$). This example motivates the definition of the $v^k(s,r)$ as a k step relevance matrix

$$
\begin{aligned}
v^1(s,r) &= v(s,r) \\
v^k(s,r) &= Max(v^{k-1}(s,r), v^{k-1}(s,s_1) \cdot v^{k-1}(s_1, r), \\
&\qquad v^{k-1}(s,s_2) \cdot v^{k-1}(s_2, r), ..., v^{k-1}(s,s_n) \cdot v^{k-1}(s_n, r))
\end{aligned}
\tag{3}
$$

As shown in the equation, we use the product as combination function and the maximum to aggregate the results. Note that, taking transitivity into account appears to be in conflict with the general idea behind our plan dimension heuristics (several refinement steps implicitly consider higher plan lengths). However, the way in which relevance is propagated (product) helps to penalize longer plans as the final relevance value is decreased (or, exceptionally, maintained if the relevance is 1) after each product.

Table 3 shows a relevance matrix for this example. In that case, $v(C_4, C_1) = 0.1$, but $v^2(C_4, C_1) = v(C_4, C_3) * v(C_3, C_1) = 0.8 * 0.7 = 0.56$.

Note that the higher the value of k the better the estimation of the relevance of service classes. The refinement of the relevance matrix is repeated until it converges (i.e. $p^{k+1}(s,r) = p^k(s,r)$) or until a timeout is received. The elevated time complexity of $O(n^3)$ for each refinement step is attenuated by the *anytime properties* of the approximation algorithm. Furthermore, recall that the number of classes n is supposed to be fixed and not overly high. Finally, note that several updates and refinements can be combined into a "batch" to be executed altogether when the system's workload is low.

Bootstrapping. There are several ways of obtaining the initial relevance matrix:

a) If there are historical records of plans they can be used to calculate the matrix.
b) An a priori distribution can be assigned using expert (heuristic) knowledge.
c) The planner can work for a while without filtering services until the number of plans generated is considered representative enough. Then the relevance matrix is calculated and refined.

Table 3. Example of relevance matrix

v		Requests							
		C_1	C_2	C_3	C_4	C_5	C_6	...	C_n
	C_1	0.9							
	C_2	0.8	1						
	C_3	0.7							
Services in	C_4	0.1		0.8					
directory	C_5	0.5							
	C_6	0.6							
	...	...							
	C_n								

These options can also be combined. For instance, the heuristic a priori distribution can be combined with the historical database as the starting matrix. In addition, if that matrix is supposed to be insufficiently informed, then it may be completed with option c).

2.3 Service Relevance Calculus

In this subsection we describe how the relevance of a service S for a request R is calculated using the relevance matrix v.

The first step to calculate the *relevance* of a service s for a request r is the mapping of both to *classes of services*. Then, the relevance between the classes is calculated. We will use the following notation:

$v(s, r)$: relevance of *class* s for the *class* r in the request, and

$V(S, R)$: relevance of service S for the service request R

Depending on whether the services are mapped to one or several classes we apply the following.

1. The simplest case is a request R that only includes a class (r) in its description. Two cases are possible:
   (a) The service S only belongs to one class (s): in this case $V(S, R) = v(s, r)$
   (b) The service S belongs to several classes $(s_1, s_2, ..., s_n)$: in this case we take the highest relevance of the different classes, i.e.
   $V(S, R) = \max(v(s_1, r), v(s_2, r), ..., v(s_n, r))$. Again, although more sophisticated functions are conceivable, we use the maximum for aggregation as it is easy to compute and intuitive for humans.

2. The request specifies a logical expression containing several classes of services $(r_1, r_2, ..., r_m)$. Now, again we have two cases:
   (a) The service S only belongs to one class (s). We evaluate logical formulas using the *maximum* for disjunctions and the *minimum* for conjunctions. For example, if the request R includes the formula $r_1 \lor (r_2 \land r_3)$, then $V(S, R) = \max(v(s, r_1), \min(v(s, r_2), v(s, r_3)))$.

(b) The service S belongs to several classes $(s_1, s_2, ..., s_n)$. In this case we combine the two previous options: the request formula is evaluated by decomposing it as in 2a); inside the expression the *maximum* is used to aggregate the service classes specified by the provider. For instance, if in the last example the service S belonged to the classes s_1 and s_2, the calculus would be:

$$V(S, R) = \max[\max(v(s_1, r_1), v(s_2, r_1)), \min(\max(v(s_1, r_2), v(s_2, r_2)), \max(v(s_1, r_3), v(s_2, r_3)))]$$

Figure 3 shows the algorithm to calculate the relevance of a service S for a request R. This is done by the *ServiceRelevance* function. This function uses the *SingleRelevance* function, which returns the relevance of the service advertisement S for one single request's class r. As described before, the request may not only include a *class of service* but also an expression (a disjunction of conjunction of classes). The two loops decompose that expression, using the minimum as combination function for the values in a conjunction and the maximum for disjunctions. Assuming that the maximum number of classes that a service can belong to is negligible, the time complexity of the algorithm is linear in the number of literals in the query.

```
ServiceRelevance(S: service advertisement; R:service request;
                 v: relevance matrix)
{
  rel = 0
  FOR ALL ConjunctionExpr IN R {
    rel' = inf
    FOR ALL r IN ConjunctionExpr {
      rel = min(rel',SingleRelevance(r,S,v))
    }
    rel = max(rel, rel')
  }
  return rel
}
SingleRelevance(r: request class; S: service advertisement;
                v: relevance matrix)
{
  rel = 0
  FOR ALL s IN S {
    rel = max(rel,v(s,r))
  }
  return rel
}
```

Fig. 3. Service relevance algorithm

2.4   Types of Service Composition Filters

When a service request is analysed by our filter, the set of services are first ranked by an estimation of the relevance of the service class for that request. Then, only the services belonging to the best ranked classes are passed on to the planner.

In order to determine the concrete services that pass the filter we consider three major options:

a) To establish a *threshold* and filter out those services whose classes have a degree of relevance lower than that threshold.

b) To return the estimated k *best* services based on the relevance of their corresponding classes. In this case the number of services that pass the filter is pre-determined.

c) To return a *percentage* of the original set of services (based on the relevance of their corresponding classes). In this case the number of services considered in the planning process depends on the directory size.

When designing the algorithms corresponding to these filters configurations, an additional problem needs to be taken into account. Services with low (or even zero) relevance values would never be considered for planning, so they could never be part of a plan (composite service), remaining with low relevance forever. This is obviously too restrictive, as our relevance values are only estimations based on the information available at some point in time. To overcome this we allow some services to be fed into the planner even though they are not supposed to be relevant enough according to the filter policy. Those additional services are chosen randomly. This random option is combined with the three aforementioned filter types to allow for an exploration of the service (class) space.

3   Role-Based Filtering

In the following we outline how our filtering framework can be effectively applied to service-oriented MAS. We show how the organisational model underlying MAS can be used as a source for the information that filters rely on. In particular, in line with common agent-oriented design methodologies [20], we use particular parts of such models, namely roles and interactions, as the key abstraction mechanisms for our approach [15,18].

To illustrate our method, we will rely on the description of the work done within the CASCOM project [9]. CASCOM aims at providing business application services, provided by intelligent agents, in dynamically changing contexts of open, large-scale, and pervasive environments. It puts forward a service-oriented MAS architecture for the medical emergency assistance domain that relies on OWL-S service descriptions.

In many service-oriented systems agents act as mere wrappers for web services. In this case, the difference between a web service and a service provided by an agent boils down to a matter of interface: agents provide access to an implemented web service by a process of wrapping the service within an ACL interface so that its invocation can be solicited by other agents through an adequate (*request*) message. However, in MAS agents should not only be able to execute a service but can also to engage in different *types of interaction* with that service, in the course of which they may play several *roles*. For example, in a medical emergency assistance scenario, an agent providing a *second opinion* service should not only be able to provide a diagnostic; it may also be required to explain it, give more details, recommend a treatment, etc. Therefore, a service provider may need to engage in several different interactions during the provision of a service.

In the next subsection, we will use the CASCOM example to outline how an ontology of the types of interactions that unfold within a service-oriented MAS can be obtained. We then show how these ontologies can be used to build up role-based service composition filters. The CASCOM example is used again to sketch how such a component can be integrated with an AI planner into a service composition planning agent.

3.1 Use Case Analysis

Setting out from a subset of the RICA organisational model described in [15], we first analyse different use cases of the application domain scenario. For each use case, we identify the types of social interaction as well as the roles (usually two) that take part in that interaction. The next step is an abstraction process in which the social (domain) roles/interactions are generalised into communicative roles/interactions.

Consider a second opinion use case in the healthcare domain [3]. In this scenario, the patient (or the physician of a local emergency centre) can ask an external health professional for a diagnosis on the basis of the symptoms and the medical records of the patient, like exams and past diseases.

A typical "conversation" between the patient and the health professional can be modelled by a sequence of (communicative) actions between the two agents involved. The patient *asks* the health professional for an opinion, providing the symptoms and the medical records. If there is insufficient information, the health professional *requests* additional information (possibly several times) and finally gives his *advisement*. If the provided diagnosis is not sufficiently clear, the patient can also solicit an *explanation*.

Starting from this conversation we can isolate 3 different interactions: (i) the second opinion exchange, which can comprise (ii) a detailed information exchange. When the second opinion exchange finishes, an explanation (iii) can occur.

The next step is an abstraction process in which, for instance, the role *SecondOpinionRequestee* can be generalized into a *MedicalAdvisor* role, which in turn can be generalized into an *Advisor* role. Similarly, the *SecondOpinion* interaction can be generalized in a *MedicalAdvisement* interaction and then in an *Advisement* interaction, in which the Advisor informs the Advisee about his beliefs with the aim of persuading the Advisee of the adequacy of these beliefs. With this process the taxonomy is refined to become more generic, so that the concepts that are individuated can be reused for other scenarios and application domains. The result of this analysis is a basic ontology of roles.

From the use case scenarios, we have derived an ontology that contains a taxonomy of types of interactions, and a taxonomy of roles that take part in those interactions [7].

3.2 Role-Based Filter

The class-based semantic service composition filter described in the previous section can be refined based on the information provided by the role and interaction ontologies. The idea is to relate roles searched in the query to roles played by agents in the composite service, that is, the roles typically involved in a plan when a role r is included in the query. For example, it is common that a *medical assistance* service includes *travel arrangement*, *arrival notification*, *hospital log-in*, *medical information exchange* and *second opinion* interactions.

Following the CASCOM approach, we use the OWL-S service parameter field to annotate service descriptions with information related to the roles and interactions in which the service provider agent can engage. Each service provider can advertise a set of possible roles from the role ontology that it can play. Similarly, in service requests it is allowed to specify the roles searched from the role ontology as a logical expression in disjunctive normal form. In [3], the role and interaction annotation procedure for service advertisements and service requests is explained in further detail. Still, for the purpose of this paper, we just need to map each role from the ontology to a *service class* of the filtering framework.

In our role based modelling approach, we use a role taxonomy that is supposed to be static over significant amounts of time. Still, the ontology *can* be extended to include new roles and types of interaction not considered before. In that case, the relevance matrix is updated with new rows and columns for those new roles. The relevance values for those new roles are unknown initially, but this can be overcome by randomly including some services with low relevance (as outlined in section 2.4) and, in general, by applying the bootstrapping techniques described in section 2.2.

3.3   Service Composition Planning Agent

We have implemented our role-based filter mechanism in JAVA on top of a P2P extension of the JADE agent platform [5]. Together with OWLS-XPlan [13], a heuristic hybrid search AI planner for the composition of OWL-S services (based on the well-known FF-planner [10]), it is part of the Service Composition Planning Agent (SCPA), a key part of the CASCOM abstract architecture [4]. Figure 4 shows the architecture and context of the SCPA.

SCPAs are capable of creating value added composite services that match specific service specifications. Once SCPAs receive service specifications from Personal Agents (PAs), they ask Service Discovery Agents (SDAs) to look for existing services in a given domain, constrained to the current context, and plan a composite value added service matching the received service specification. The generated plans are forwarded to Service Execution Agents (SEA), which manage the execution of the composite services. In addition, the plan and the original request are passed on to the filter component to update the historical information about plans.

The performance of the CASCOM approach in general, and the adequacy of the SCPA composition planning approach based on an AI planner and configurable, adaptive, service class based composition filters in particular, will be evaluated during a field trial in the medical emergency assistance domain.

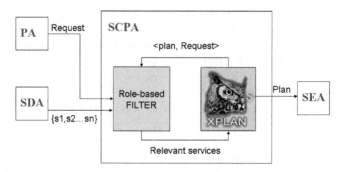

Fig. 4. SCPA architecture

4   Conclusions

In this paper we have presented a filtering method for semantic service composition. This approach enables us to effectively apply recent AI planning techniques to semantic service composition by reducing the (potentially huge) set of input services that are passed on to the planner. For this purpose, we make use of service class information, so as to cluster services based on certain properties. We have argued that how this information can be obtained from the organisational structure underlying service-oriented MAS. In particular, based on real-world use cases of the CASCOM project, we have derived role and interaction ontologies that are used for the efficient caching and processing of plan information. Finally we sketched how such a role-based filter component can be integrated into the architecture of the CASCOM service-oriented MAS.

There is currently a limited number of composition planning approaches based on OWL-S (e.g. [8,16,17,19]). However, they are geared towards standard service-oriented architectures. To the best of our knowledge none of them makes use of organisational information within a service-oriented MAS to filter services.

In the near future we will perform a more quantitative evaluation of this approach as part of the CASCOM field trial. We will also compare the performance of our role-based method to other semantic service composition filter instantiations. In particular, we will look into the use of OWL-S Service Category information for determining the service class information required by the filter. Future work will investigate how to incorporate quality of service and trust mechanisms [1] into our approach.

References

1. Billhardt, H., Hemoso, R., Ossowski, S., Centeno, R.: Trust-based Service Provider Selection in Open Environments. Proceedings of the 2007 ACM Symposium on Applied Computing (SAC), Seoul, Korea (2007) 11-15
2. Cáceres, C., Fernández, A., Ossowski, S.: CASCOM - Context-aware Health-Care Service Coordination in Mobile Computing Environments. ERCIM News **60** (2005) 77-78

3. Cáceres, C., Fernández, A., Ossowski, S., Vasirani, M.: Agent-Based Semantic Service Discovery for Healthcare: An Organizational Approach, IEEE Intelligent Systems **21** (6) (2006) 11-20
4. Cáceres, C., Fernández, A., Ossowski, S., Vasirani, M.: An abstract architecture for semantic service coordination in agent-based intelligent peer-to-peer environments. The 20th ACM 2006 Annual Symposium on Applied Computing (SAC). Dijon (France) (2006)
5. CASCOM Consortium.: CASCOM Project Deliverable D4.1: IP2P Network Architecture. (2006)
6. Decker, K., Sycara, K., Williamson, M.: Middle-agents for the internet. In International Joint Conference on Artificial Intelligence. Nagoya, Japan (1997)
7. Fernandez, A., Vasirani, M., Caceres, C., Ossowski, S.: A Role-based Support Mechanism for Service Description and Discovery. In: J.Huang et al. (eds.): Service-Oriented Computing: Agents, Semantics, and Engineering. Lecture Notes in Computer Science. Springer-Verlag (2007)
8. Hamadi, R., Benatallah, B.: A Petri-Net-Based Model for Web Service Composition. Proc. 14th Australasian Database Conf. Database Technologies, ACM Press (2003) 191-200
9. Helin, H., Klusch, M., Lopes, A., Fernandez, A., Schumacher, M., Schuldt, H., Bergenti, F., Kinnunen, A.: Context-aware Business Application Service Coordination in Mobile Computing Environments. In AAMAS05 workshop on Ambient Intelligence - Agents for Ubiquitous Computing. Utrecht, The Netherlands (2005)
10. Hoffmann, J., Nebel, B.: The FF Planning System: Fast Plan Generation through Heuristic Search. Journal of Artificial Intelligence Research (JAIR) **14** (2001) 253-302
11. Huhns, M. N., Singh, M. P.: Service-Oriented Computing. John Wiley & Sons (2005)
12. Huhns, M. N., et al.: Research Directions for Service-Oriented Multiagent Systems. IEEE Internet Computing **9** (6) (2005)
13. Klusch, M., Gerber, A., Schmidt, M.: Semantic Web Service Composition Planning with OWLS-XPlan. Proceedings 1st Intl. AAAI Fall Symposium on Agents and the Semantic Web. Arlington VA, USA (2005)
14. Peer, J.: Web service composition as AI planning - a survey. Technical report, Univ. of St. Gallen, Switzerland (2005)
15. Serrano, J.M., Ossowski, S.: A computational framework for the specification and enactment of interaction protocols in multiagent organizations. Journal of Web Intelligence and Agent Systems, Idea Press (to appear)
16. M. Sheshagiri, M. desJardins, T. Finin.: A planner for composing services described in DAML-S. Proceedings of AAMAS 2003 Workshop on Web Services and Agent-Based Engineering (2003)
17. S. Tarkoma, M. Laukkanen.: Adaptive agent-based service composition for wireless terminals. Proceedings of Seventh International Workshop on Cooperative Information Agents (M. Klusch et al, eds.). Helsinki, Finland. Springer Verlag, LNAI 2782 (2003) 16-29
18. Partsakoulakis, I., Vourus, G.: Roles in MAS. An Application Science for Multiagent Systems (A. Wagner, ed.). Kluwer (2004) 133-155
19. D. Wu, B. Parsia, E. Sirin, J. Hendler, D. Nau.: Automating DAML-S web services composition using SHOP2. Proceedings of the 2nd International Semantic Web Conference (ISWC2003). Sanibel Island, Florida, USA (2003) 20-23
20. Zambonelli, F., Jennings, N. R., Wooldridge, M.: Organizational Abstractions for the Analysis and Design of Multi-agent Systems. Agent-Oriented Software Engineering: First International Workshop, AOSE 2000, Limerick, Ireland (2000) 235-251.

Towards a Mapping from BPMN to Agents

Holger Endert, Benjamin Hirsch, Tobias Küster, and Sahin Albayrak

DAI-Labor, Technische Universität Berlin
{holger.endert,benjamin.hirsch,
tobias.kuester,sahin.albayrak}@dai-labor.de

Abstract. In industry, people who design business processes are often different from those designing the technical realization. Also, they generally use different languages, such as BPMN on the one hand and UML on the other. While agents are theoretically suitable for designing and implementing business ideas, multi-agent methodologies are generally not geared towards them.

In this paper, we describe the first step of mapping business process diagrams to agent concepts. To this end, we present a graph based representation of BPMN together with structural and semantical analysis methods. These provide the necessary formal grounding for the mapping we have in mind.

1 Motivation

The last couple of years have seen an increasing interest in service oriented architectures (SOA) and business process modelling. While SOA is certainly a hype within industry, there are a number of features that have been researched within the agent community for many years. On the other hand there are quite a few things that the agent community can learn from SOA in general. One of these in our view is the focus on process models, and the use of graphical notations. The graphical design of workflow and interaction protocols provides highly desirable advantages. An intuitive representation enables a broad range of persons to become a clear understanding, what a diagrams objective is, without being familiar with a specialized definition language. On the other hand, graphical languages often lack in providing a defined semantic. Any use, which relies on a distinct meaning of a particular diagram, e.g. the mapping to formal languages, code generation or validation, is difficult or even impossible.

Our aim is to take the good things from SOA related research and adapt them to agents. In particular, we want to map graphical notations to agent systems. In order to do this however, we need a firm formal basis for the graphical notation. We have chosen the Business Process Modeling Notation (BPMN), which is intended to define business processes. In industry as well as academic research this language is of increasing interest. One reason for that is its simple, but expressive set of notations. This makes it adequate for non-experts equipped with domain knowledge to design business processes within their domain. Although there are other well studied graphical languages which have a formal semantics, they are often not the first choice for system or process design.

J. Huang et al. (Eds.): SOCASE 2007, LNCS 4504, pp. 92–106, 2007.
© Springer-Verlag Berlin Heidelberg 2007

However, up to now, the semantic of BPMN is not formally defined. Therefore, this work describes the first step towards a mapping from BPMN to agents, as it addresses the topic of defining a formal syntax and semantic of BPMN with respect to business models. On top of this, validation of diagrams is presented, which can handle inconsistencies in both, syntax and semantic. The validation is grounded on well studied theories for petri nets, such that we can apply the theoretical results given there. As a result, we obtain a language, that is easy to understand and usable within a broad spectrum of applications. Future work will describe a mapping of the normalized BPMN diagrams to our agent framework.

This paper is structured as follows: Section 2 provides some informal background on the BPMN language. In Section 3, the BPMN is formally defined as graph structure, a normal form is presented, and a transformation to a corresponding petri net variant is shown. At the end, we discuss properties and analysis methods. We conclude this paper with a comparison of similar work (Section 4) and some final remarks (Section 5).

2 BPMN

The *Business Process Modeling Notation* specification [1], which is maintained by the Object Management Group, is a graphical notation for describing various kinds of processes. The main notational elements in BPMN are *FlowObjects*, that are contained in *Pools* and connected via *Sequence-* and *MessageFlows*. They subdivide in *Events*, atomic and composite *Activities* and *Gateways* for forking and joining. SequenceFlows describe the sequence in which the several FlowObjects have to be completed, while MessageFlows describe the exchange of messages between Pools. Thus, BPMN combines the definition of local workflows and interaction protocols between them.

An example diagram is shown in Figure 1. It models the interaction between a patient and a doctor (adapted from [1]). Here, the patient requests a prescription, which he receives in the next task. This is modelled as an activity (rounded rectangle) with an outgoing message (dashed arrow). The doctors workflow includes the corresponding reception of the request, whereafter it begins to do two things in parallel (the gateway-diamonds enclose the parallel activities). In the first branch, the prescription is filled out and send to the patient, in the other some accounting tasks are executed, which are contained in a subprocess. Finally, both processes are enclosed in *Start-* and *End-*Events (circles).

As can be seen, the meaning of the diagram can be understood intuitively and without any specific knowledge about the semantics of BPMN by human. This strength is relativized in that BPMN is more a graphical notation than a formal one, which is required for automatic interpretation. Although the specification includes some semantics and even defines a mapping to WS-BPEL (see Chapter 11 of the specification [1]), much of the semantics is especially tailored for this mapping and the focus of the specification is clearly the visual representation.

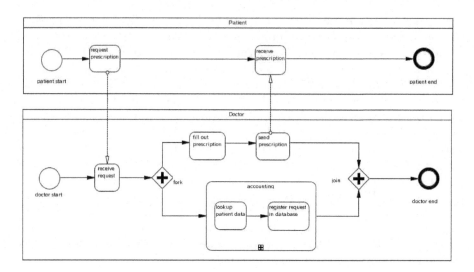

Fig. 1. BPMN example diagram

3   Validation of BPMN

Because of BPMN's visual nature, it allows to define simple interactions quite easy, but for sophisticated and complex designs, additional tool support is indispensable. Therefore we propose a validation process for BPMN diagrams (BPDs), which consists of a number of steps. First, diagrams are verified to satisfy certain structural conditions. To this end, a *normal form* is introduced in the next Subsection, which allows for checks of structural correctness. Thereafter, we propose a semantic analysis of the normalized diagrams based on petri nets, which will locate inconsistencies in a defined workflow and protocol in Subsection 3.2. Finally, we discuss the results which were obtained from applying the petri net approach in Subsection 3.3.

3.1   Normalized Business Process Diagrams

Because of the variety of semantically equivalent syntax of BPMN constructs, we define a *normal form* of BPDs. This is an extension of the approach taken in [2], which defines a subset of the complete language as a *Core BPD*. However, for the scope of this work, which includes the multi-agent case as well, this definition is too restrictive, because it is limited to a single process. In the first instance, a Business Process Diagram is defined as the following graph structure:

Definition 1. *(BPD-Graph) - Let $BPD = (O, F, src, tar)$ be a graph with*

- *O — the set of nodes (objects) in the BPD-Graph.*
- *F — the set of edges (message and sequence flows) in the BPD-Graph.*
- *$src, tar : F \rightarrow O$ two functions, which identify the source and target objects of each edge.*

In order to define the structure of a normalized BPD-Graph, some additional notations are required for the nodes, edges and relations of the graph. Let O be partitioned into the disjoint subsets O^E, O^A, O^G, O^P, where

- O^E — the set of event-nodes, which can be further partitioned into the disjoint subsets O_S^E, O_E^E, O_I^E, i.e. start-, end- and intermediate events.
- O^A — the set of activity-nodes, which can be further partitioned into the disjoint subsets O_{At}^A, O_{Sub}^A, i.e. the atomic activity nodes and the subprocess nodes[1].
- O^G — the set of gateway-nodes, which can be further partitioned into the disjoint subsets O_S^G and O_M^G, i.e. the splitting and the merging gateway nodes. These can again be partitioned into the subsets for exclusive (XOR), inclusive (OR) and parallel (AND) split and merge gateways ($O_{S,X}^G, O_{S,O}^G, O_{S,A}^G$, $O_{M,X}^G, O_{M,O}^G$ and $O_{M,A}^G$).
- O^P — the set of pool-nodes.

Next, let F be partitioned into the two disjoint subsets F^S and F^M, i.e. the set of sequence and message flows. The functions src^S, src^M, tar^S and tar^M are defined equal to the src and tar functions, but restricted to given domains F^S and F^M. Finally, the following functions are given:

- $parent : O \to O$, where $parent(o)$ returns the object in which o is located in. If $o \in O^P$, it is defined as identity function.
- $contains : O \to \{O\}$, where $contains(o)$ retrieves the set of contained nodes. If $o \notin O^P$, it holds that $o \in contains(parent(o))$.
- $pool : O \to O^P$, which returns the pool (the most top-level element of a workflow), in which a node is contained. This function is defined recursively with the $parent$ function.

Now the *Normalized BPD-Graph* is defined as follows:

Definition 2. *(Normalized BPD-Graph) - A BPD-Graph is said to be normalized, iff it satisfies the following conditions:*

- $\forall f \in F^S : tar_S(f) \notin O_S^E$, *i.e. start elements have no ingoing sequence edges.*
- $\forall f \in F^S : src_S(f) \notin O_E^E$, *i.e. end elements have no outgoing sequence edges.*
- $\forall o_1, o_2 \in O_S^E : parent(o_1) = parent(o_2) \Rightarrow o_1 = o_2$, *i.e. for each (sub-) process, there exists a unique start node.*
- $\forall o_1, o_2 \in O_E^E : parent(o_1) = parent(o_2) \Rightarrow o_1 = o_2$, *i.e. for each (sub-) process, there exists a unique end node.*
- $\forall o \in O \backslash (O_E^E \cup O^P), \exists f \in F^S : src^S(f) = o$, *i.e. each node (except end nodes and pools) has an outgoing sequence.*
- $\forall o \in O \backslash (O_S^E \cup O^P), \exists f \in F^S : tar^S(f) = o$, *i.e. each node (except start nodes and pools) has an ingoing sequence.*

---

[1] Atomic activities are all those which do not contain any other nodes directly. This holds also for reference sub-processes.

- $\forall o \in (O^A \cup O_E^E \cup O_S^G \cup O_I^E), \exists! f \in F : tar^S(f) = o$, i.e. each node of type activity, end-event, intermediate-event or split-gateway has exactly one ingoing sequence.
- $\forall o \in (O^A \cup O_S^E \cup O_M^G \cup O_I^E), \exists! f \in F : src^S(f) = o$, i.e. each node of type activity, start-event, intermediate-event or merge-gateway has exactly one outgoing sequence.
- $\forall f \in F : src(f) \neq tar(f)$, i.e. all edges (message- and sequence flows) do not connect an object with itself.
- $\forall f \in F^M : pool(src^M(f)) \neq pool(tar^M(f))$. This statement means, that messages are only allowed between objects, which are located in different pools.
- $\forall f \in F^M : src^M(f) \in O_{At}^A \wedge tar^M(f) \in O_{At}^A$, i.e. message flow is only allowed between atomic activities.
- $\forall f \in F^S : src^S(f) \in contains(parent(tar^S(f))) \wedge tar^S(f) \in contains (parent(src^S(f)))$ i.e. sequence flow is only allowed within the same container (pool, subprocess). Note that this definition forbids sequence flows between pools, because parent is defined as identity function for pools.

Thus, a *normalized BPD* fulfills simple structural conditions. These can be evaluated using a set of relatively simple rules. Furthermore, we define a number of rules that detect syntactically incorrect diagrams. Therefore we define two sets of graph transformation rules[2], one for applying a normalization, that does not change the semantics of the diagram, and the other for detecting errors.

Definition 3. *(Normalization Rules: R_{Norm}) - The normalization rules are defined as follows:*

1. *AddStartEvents: If any $o \in (O \backslash O_S^E)$ does not have an incoming sequence flow, a start event is added, and connected to the object.*
2. *AddEndEvents: If any $o \in (O \backslash O_E^E)$ does not have an outgoing sequence flow, an end event is added, and connected to the object.*
3. *AddSplitGateway: If any $o \in O$ has more than one outgoing sequence flow, a parallel split gateway is added, and the sequence flow is changed accordingly.*
4. *AddMergeGateway: If any $o \in O$ has more than one incoming sequence flow, a parallel merge gateway is added, and the sequence flow is changed accordingly.*
5. *UniqueStartNode: For each $o \in O$, if $|O_S^E \cap contains(o)| \geq 2$, i.e. there is more than one start node, they are replaced by a unique start node and a parallel split gateway.*
6. *UniqueEndNode: For each $o \in O$, if $|O_E^E \cap contains(o)| \geq 2$, i.e. there is more than one end node, they are replaced by a unique end node and a parallel merge gateway.*
7. *MessageSrc: For each $f \in F^M$, if $src^M(f) \in O_{Sub}^A$, i.e. its source node is a subprocess, a send-activity is added within its process between the start and the following node, and the message source is connected to that.*

---

[2] We use typed attributed graph transformations with node type inheritance and negative application conditions. These are applied to the source graph as long as possible, until a result graph is obtained.

8. *MessageTar: For each $f \in F^M$, if $tar^M(f) \in O^A_{Sub}$, i.e. its target node is a subprocess, a receive-activity is added within its process between the end- and the previous node, and the message source is connected to that.*

Generally, the normalization rules are used to identify BPMN syntax, which may be correct but doesn't conform to the normalized structure, and replaces them with elements conform to the normal form. The rules R_{Error} for detecting errors are also straightforward. Each condition in Definition 2 is negated and implemented in a single rule, and if these are applicable, error-objects are connected to them. For this purpose, the BPD-Graph is extended as follows:

Definition 4. *(Extended BPD-Graph) - Let E-BPD be a graph with $E-BPD = BPD \cup (Err, ErrLink, src_X, tar_X)$, where*

- *Err is a distinguished set of error objects.*
- *ErrLink is a distinguished set of error edges.*
- *$src_X : ErrLink \to Err$, i.e. is a function that associates an error edge with an error.*
- *$tar_X : ErrLink \to O$, i.e. is a function that associates an error edge with an object of a BPD-Graph.*

Definition 5. *(Structurally Correct BPD-Graph) - Let **ebpd** be an Extended BPD-Graph with $Err = ErrLink = \oslash$. Let further R_{Norm} and R_{Error} be the set of normalization and error annotation rules as defined above.*
*A BPD-Graph **bpd** is said to be structurally correct, if $**bpd** = **ebpd**|_{BPD}$, where $**ebpd**|_{BPD}$ denotes the BPD-Graph that results from omitting all extensions of Definition 4, and for $ebpd_t = R_{Error}(R_{Norm}(ebpd))$ holds $Err_t = ErrLink_t = \oslash$, i.e. after applying the rules for normalization and error detection, no error elements are created[3].*

3.2 Extending BPMN to Petri Nets

Although structural correctness allows to detect simple errors, it is not a sufficient condition for correct workflow and protocols. For analyzing these properties, additional verifications using graph transformations can be done. We want to ensure that both, the workflow and the protocol, do not contain any inconsistencies in message- or sequence flow (e.g. sent messages will never be received). In order to address these issues, we extend BPDs such that transformations similar to the token game of petri nets are applicable, which are well understood. Using petri nets analysis, we are able to detect cycles, deadlocks, and reachability properties of BPMN diagrams and their elements. Before presenting the transformation rules, we have to consider some issues regarding the structure of BPMN diagrams. These are different from that of petri nets in that on the one hand they contain hierarchies, and on the other they have different types of transitions (message- and sequence flow). The former property is dealt with by

---

[3] $R_2(R_1(G))$ denotes the sequential application of both rule sets, each as long as possible.

flattening the net, i.e. by resolving hierarchies. The latter is addressed by using an extension of regular petri nets, which allows to distinguish two token types [3]. Taking this into consideration, we define a PNet BPD-Graph in terms of *place transition nets* as follows.

Definition 6. *(PNet BPD-Graph) - A PNet BPD-Graph is an extended petri net with PNet BPD = (P, T, V, C, W) with*

- *P — the set of places of the net.*
- *T — the set of transitions of the net.*
- *V — the set of arcs, connecting places and transitions (flow relation), with $V \subseteq (P \times T) \cup (T \times P)$.*
- *C — the capacity of tokens for each place is a tuple (C_S, C_M), which defines the maximum number of tokens per type (sequence and message tokens) that are permitted.*
- *$W : (P \times T) \cup (T \times P) \rightarrow (\mathbb{N} \times \mathbb{N})$ the weight function, which defines a weight tuple (W_S, W_M) to each existing place-transition connection. The components of the tuple refer to the weights for message- and sequence flows.*

For a given PNet BPD-Graph, there exists a current marking, called **a**, which holds the tokens for each place in a given state of the net. We subsequently use the following functions to refer to the markings of given places p:

- *$a : P \rightarrow \mathbb{N} \times \mathbb{N}$ — returns the token tuple of a place.*
- *$a_S : P \rightarrow \mathbb{N}$ — the number of sequence tokens, which are present in marking a for p.*
- *$a_M : P \rightarrow \mathbb{N}$ — the number of message tokens, which are present in marking a for p.*

Beside the extended token representation, this definition conforms to the usual petri nets, and hence its formal semantics can be used here. Next, we define the construction of a PNet BPD Graph out of a structurally correct and normalized BPD-Graph:

Definition 7. *(Constructed PNet BPD-Graph) - Let **bpd** be a structurally correct BPD-Graph. Then the construction of a corresponding PNet BPD-Graph is defined as follows:*

- *The set of places P is composed of the set $O^E \cup O^A_{At} \cup O^G$, i.e. the set of events, atomic activities and gateways.*
- *The set of transitions T is defined as follows:*
  - *For each $f \in F^S$ with $\nexists f_2 : f \neq f_2 \wedge (src(f) = src(f_2) \vee tar(f) = tar(f_2))$, one transition is created. These are called **simple sequence transition**.*
  - *For each $f \in F^M$ one transition is created. These are called **simple message transition**.*
  - *For each $g \in (O^G_{S,A} \cup O^G_{M,A})$ one transition is created. These are called **complex parallel transition**.*

- *Let $g \in (O_{S,X}^G \cup O_{M,X}^G)$ be an exclusive gateway. Let further P_{out} be the set of nodes, which are reachable via sequence flow from g, i.e. $\forall p \in P_{out}$ $\exists f \in F^S : src(f) = g \wedge tar(f) = p$, and P_{in} the set of nodes, from which g is reachable. If $g \in O_S^G$, then for each element of P_{out} one transition is created, otherwise one one transition for each element in P_{in} is created. These are called **complex exclusive transition**.*
- *Let $g \in (O_{S,O}^G \cup O_{M,O}^G)$ be an inclusive gateway. Let P_{in} and P_{out} be defined as above. If $g \in O_S^G$, then one transition for each element of the power set $\wp(P_{out})$ is created. Otherwise, if $g \in O_M^G$, then one transition for each element of the power set $\wp(P_{in})$ is created. These are called **complex inclusive transitions**.*
- *The flow relation V is defined as follows. Given these functions*

$$start, end \; : \; O \to O$$
$$start(o) = if \quad o \notin (O_{Sub}^A \cup O^P)$$
$$then \quad id(o)$$
$$else \quad !(contains(o) \cap O_S^E)$$
$$end(o) = if \quad o \notin (O_{Sub}^A \cup O^P)$$
$$then \quad id(o)$$
$$else \quad !(contains(o) \cap O_E^E)$$

where $!(S)$ refers to an arbitrary element of a set. Because of the uniqueness of start and end events, these functions are well defined.

- *For each simple (message and sequence) transition t, the relations $(end(src(f)), t)$ and $(t, start(tar(f)))$ is in V, where f is the sequence- or message flow, that lead to the construction of t.*
- *For a complex parallel transition t, let g be the gateway from which it was created. If $g \in O_S^G$, then the relation (g, t) and for each sequence flow f_{out} with $src(f_{out}) = g$ the relations $(t, start(tar(f_{out})))$ belong to V. Otherwise, if $g \in O_M^G$, then the relation (t, g) and for each sequence flow f_{in} with $tar(f_{in}) = g$ the relation $(end(src(f_{in})), t)$ belong to V.*
- *Let $t_1, .., t_n$ be the set of complex exclusive transitions, which were created from g. If $g \in O_S^G$, then the relations (g, t_i) for $1 \le i \le n$ are in V, and for each t_i, let p_i be the element of P_{out} that lead to the construction of t_i. Then, the relations $(t_i, start(p_i))$ belong to V as well. Otherwise, if $g \in O_M^G$, then the relations (t_i, g) for $1 \le i \le n$ are in V, and for each t_i, let p_i be the element of P_{in} that lead to the construction of t_i. Then, the relations $(end(p_i), t_i)$ belong to V as well.*
- *Let $t_1, .., t_n$ be the set of complex inclusive transitions, which were created from g. If $g \in O_S^G$, then the relations (g, t_i) for $1 \le i \le n$ are in V, and for each t_i, let P_i be the set of elements of $\wp(P_{out})$ that lead to the construction of t_i. Then for each $p_i \in P_i$ the relation $(t_i, start(p_i))$ belongs to V as well. Otherwise, if $g \in O_M^G$, then the relations (t_i, g) for $1 \le i \le n$ are in V, and for each t_i, let P_i be the set of elements*

of $\wp(P_{in})$ that lead to the construction of t_i. Then for each $p_i \in P_i$ the relation $(end(p_i), t_i)$ belongs to V as well.
- No further relations are in V.
- The capacity C of each place is unlimited, i.e. the tuple (∞, ∞).
- The weight function is defined as follows:
  - For all simple transitions t, which were constructed from message flow, the weight is defined as

$$W((p_1, t)) = W((t, p_2)) = (0, 1)$$

  - For all other transitions t, the weight is defined as

$$W((p_1, t)) = (1, In(p_1))$$
$$W((t, p_2)) = (1, Out(p_2))$$

  where In and Out are the functions retrieving the cardinality of the set of out- and ingoing message flows of a node p respectively.
- The start marking a_0 is defined by:
  If $p \in O_S^E \wedge parent(p) \in O^P$, then $a_0(p) = (1, 0)$. For all other places a_0 is defined with $(0, 0)$.

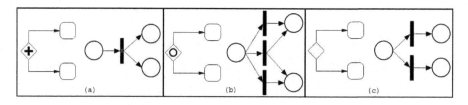

Fig. 2. Translation of Gateways to PNet BPD elements - (a) - parallel, (b) - inclusive and (c) - exclusive or, each with two outgoing sequences

As can be seen, the most complex construction rules are for gateways. In Figure 2, the mapping for the corresponding split gateways to the petri net is given. For illustrating a construction, we present it exemplarily on the BPD given in Figure 1, which is already structurally correct. The resulting PNet BPD-Graph is given in Figure 3. There, each node except the subprocess *accounting* is transformed into a place. Most transitions are simple transitions, and thus are created directly from message- or sequence flow. Exceptions are T_8 and T_{13}, which are created from the gateways. The flow relation is also defined straightforward: Whenever simple transitions are created, places, which were originally connected by the sequence or message flow, are connected to the transitions accordingly. For gateways, the constructed transitions are linked to all incoming or outgoing places respectively. The weight function is interesting for (T_1, S_2), (S_3, T_3), (S_6, T_7) and (T_9, S_9), where it is $(1, 1)$. For the receive tasks, it means that a received message is required in order to enable the following transition, and for send tasks, a message token for sending is created and subsequently available. (S_2, T_4), (T_4, S_6),

(S_9, T_5) and (T_5, S_3) are defined $(0,1)$, implementing the sending and receiving of a message. (T_8, S_{10}) and (S_{13}, T_{13}) resolve the hierarchy of the subprocess *accounting*. Finally, the start marking is $(1,0)$ for S_1 and S_5, all others are marked $(0,0)$.

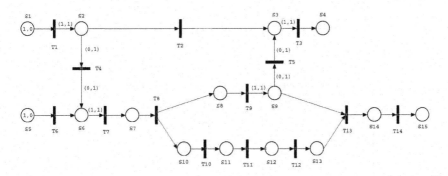

Fig. 3. The constructed PNet BPD-Graph of the BPD

3.3   Analyzing PNet BPDs

The construction of PNet BPD-Graphs comes along with some (computational) effort that should be motivated adequately. So, what are the advantages of doing so? The obvious answer is that PNet BPDs allow the use of well studied methods, which are able to detect inconsistencies. When considering the translation of BPDs into some executable language, another big advantage becomes evident. Possible inconsistencies may be located a priori, i.e. before the target language is created and executed. To clarify this, we highlight some desired properties of BPDs, and argue that these can be validated using the corresponding constructed PNet BPD and applying petri net analyzation techniques. In particular, these properties are *termination*, *relevance* (of contained elements), and *boundedness*.

The first property we discuss is also a very essential one, named *termination*. It refers to the ability of a diagram to reach a desired and final state that marks the (successful) termination of the designed workflow. Termination may never be achieved in cases were the workflow is trapped in an endless loop or deadlocks occur, such that successor states are not reachable. An example of both cases can be seen in Figure 4. Both parts are structurally correct, thus applying the error detection rules will not offer any result. But both are obviously inconsistent.

Correct termination of a BPD occurs when the workflow halts at the most top-level end element. Speaking in terms of a PNet BPD, this means that the place which is used for the top-level end event owns a token, and all others don't. Therefore, we define next the *Termination Criterium* of a PNet BPD as follows:

Definition 8. *(Termination Criterium of a PNet BPD) - Let **pbpd** be a PNet BPD-Graph, constructed from **bpd** and with start marking a_0. Let further be*

$R(pbpd, a_0)$ the set of all markings of **pbpd** reachable from a_0. Then, **pbpd** is said to satisfy the termination criterium, iff

$$\exists a_n \in R(pbpd, a_0) :$$
$$\forall p \in P :$$
$$if \quad p \in O_E^E \wedge parent(p) \in O^P$$
$$then \quad a_n(p) = (1,0)$$
$$else \quad a_n(p) = (0,0)$$

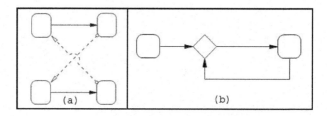

Fig. 4. Two possible diagram-parts that do not terminate: (a) A part of a BPD that does not terminate because of inconsistent message flow. (b) A part of a BPD that does not terminate because of an infinite loop.

This criterium is a necessary condition for termination, i.e. if it does not hold for a certain PNet BPD, it will never terminate. On the other hand, if this criteria is fulfilled, we can only claim that termination is probably achieved. The reason for that lies in the fact that a PNet BPD is not able to include the conditional statements which are available for each gate of a split gateway. For instance, if the only branch which leads to termination is never chosen, then in practice the workflow will never terminate.

The approach for checking this property with petri nets is known as the reachability property. By analyzing the reachability of the petri net, it is evaluated which markings are derivable through firing of transitions for a given initial marking of a net. For instance, reachability is evaluated by creating the coverability tree, which may be done using breadth first search of possible markings. If the net is bounded, the tree is also called *reachability tree*, because it contains all possible markings. Hence, this approach is suitable for our approach to check the termination criteria of a PNet BPD, which after construction already contains all necessary elements (the graph and an initial marking). An algorithm for creating the coverability tree is given in [4]. Due to the explosion of state space, this approach is limited to smaller nets, or nets with small degree of branching, because in general its computational costs grow exponentially with the size of the graph. Improvements in performance are achieved if independent parts (e.g. sub-processes) of a net are analyzed separately. But to do this, these parts must be identified beforehand, which is generally not trivial.

The next property we analyze is the relevance of both, a complete BPD and its contained elements. This is closely related to the termination property, and means, that every element (node) of a BPD becomes active in at least one particular case. If this property is not given, a diagram contains elements which are never reached, and hence are *irrelevant* for the designed workflow. We define relevance for places of PNet BPDs as follows:

Definition 9. *(Relevance of Places in a PNet BDP)* - Let **pbpd** be a PNet BPD-Graph, constructed from **bpd** and with start marking a_0. Let further be $R(pbpd, a_0)$ the set of all markings of **pbpd** reachable from a_0. Then a place $p \in P$ is said to be relevant, iff

$$\exists a_n \in R(pbpd, a_0) :$$
$$a_{n,S}(p) \geq 1$$

This means that if there exists a marking which is derivable from a_0 and contains a positive value of sequence tokens for a given place, this is *relevant* for the designed workflow. Note that we do not include message tokens for relevance, because places may become owner of a message token without being active (e.g. a message is only actively received if the node was activated by a sequence flow). On top of this definition, we can define a global relevant PNet BPD:

Definition 10. *(Global Relevance of a PNet BDP)* - Let **pbpd** be a PNet BPD-Graph, constructed from **bpd** and with start marking a_0. Let further be $R(pbpd, a_0)$ the set of all markings of **pbpd** reachable from a_0. Then **pbpd** is said to be relevant, iff

$$\forall p \in P : \exists a_n \in R(pbpd, a_0) :$$
$$a_{n,S}(p) \geq 1$$

The only difference here is that relevance is required for all places of the net. Again, the relevance property is a necessary condition. If it is not satisfied, the net contains inconsistencies. If it is, no statements can be made for sure. Definition 10 indicates already which method can be used to analyze relevance: The construction of the coverability tree allows to decide if a graph is relevant.

The last property we discuss is boundedness, which in contrast to those mentioned before defines a sufficient condition. As in ordinary petri nets, we define boundedness of places as a limiting border of tokens, which is not exceeded in any reachable marking. Within petri nets, this property becomes interesting, if tokens are used as data buffers. The presence of a bound indicates that no buffer overflow will occur. In BPDs, messages are sent asynchronous, such that buffers are required as well (this counts also very often for agents, which communicate via speech acts). Subsequently, in a PNet BPD, the number of message tokens in a place represents the actual filling of the corresponding buffer. If a bound for message tokens can be found which is never exceed, the graph can said to be *save* with respect to message buffer overflow[4]. We define the boundedness of a PNet BPD as follows:

---

[4] Note, that safety for petri nets is defined as a 1-bounded net, i.e. that a bound exists and has a value of 1

Definition 11. *(k-Boundedness of a PNet BDP) - Let* **pbpd** *be a PNet BPD-Graph, constructed from* **bpd** *and with start marking* a_0. *Let further be R(pbpd, a_0) the set of all markings of* **pbpd** *reachable from* a_0. *Then* **pbpd** *is said to be k-bounded, iff*

$$\forall p \in P, a_n \in R(pbpd, a_0) :$$
$$a_{n,M}(p) \leq k$$

Finding a bound is related to the question whether a coverability tree is also a reachability tree. If so, the bound k can be found by looking for the maximum marking of a place within the tree of reachable markings. Otherwise, if the tree is not a reachability tree, no bound exists. From this property we can infer the following: Since it is a sufficient condition, a bound means that surely no buffer overflow will occur. Therefore it will not discover inconsistencies in a graph, but ensure that this kind of error will not occur. However, if a graph is unbound, no statements can be made for the same reasons as before.

Alltogether we obtain a representation of BPMN diagrams, for which syntactic and semantic validations are possible. Additionally, we provide a set of techniques for executing validations on that graphical model, and therefore we show that some inconsistencies may be discovered. Finally we also note that we are currently working on a formal proof to show that all statements that hold for a constructed PNet BPD also hold for the BPD from which it was constructed.

4   Related Work

We subsequently give an overview of similar formal approaches and applications related to business process modelling. Somewhat close to what we have in mind is the work of Vidal et al. [5]. They use BPEL4WS[6] for describing workflows and propose to combine them with DAML-S for achieving semantical service descriptions. On top of this, agents can be used as service providers, allowing to dynamically select and execute web services. Since the BPMN specification provides a mapping to BPEL4WS, their ideas will influence our future work.

Likewise, the usage of petri nets for workflow design is studied for a while by Aalst et al. [7]. Other approaches apply a specific variant of petri nets (CPN). For instance in [8], these are directly used to model multi-agent systems.

Regarding BPMN, there are a number of tools that allow to model business processes, some of which include a (limited) translation into BPEL. Many more tools are based on BPEL to model processes that are directly executable. Ouyouang et al. [2] highlights some of the difficulties of translating BPMN into block-oriented languages, and describe a translation of BPMN to BPEL, in which they provide methods that deal with parallelism and joins. Similar work is that of Koehler et al. [9] who designed a method to remove cycles in graphs and replace them with equivalent structures.

As the importance and impact of SOA and friends starts to become clear in the area of agent technologies, a number of people have worked on different ways of incorporating knowledge and experiences of the different fields. Casella et al. [10] and Mantell [11] have worked on creating tools to translate UML diagrams to BPEL. Mantell translates UML activity diagrams to executable BPEL code, while Casella et al. start from protocol diagrams designed in the UML extension Agent-UML [12] and create abstract BPEL processes. While these approaches are similar to ours, we chose BPMN because it is more accessible to non-computer scientists than UML or an extension. Especially business people think in terms of processes rather than e.g. activity or class diagrams.

There are other approaches that directly link webservices and agency. Bozzo et al. [13] apply the BDI paradigm in order to create adaptive systems based on webservices. Here, they start from a BDI-type MAS (based on AgentSpeak [14]) and extend it to use web services as primitive actions. Walton [15] suggests to differentiate between agent interaction protocols and agent body, and therefore to allow web services to be seamlessly incorporated into MAS, and vice versa.

5   Conclusion and Future Work

In this paper, we have laid the foundation of a mapping from the business process modeling notation BPMN to agents. In order to arrive at a semantically sound translation, we have here focussed on the formal specification of a normal form for BPMN diagrams. We have provided a translation method that maps arbitrary BPMN diagrams to the normalized form, and that checks for certain structural properties. The normalized form is then transformed into a petri-net which allows for further semantic analysis.

Ultimately we want to enable domain experts to design workflows and agent behaviour in a language they understand. We currently work on a mapping from BPMN to the agent framework JIAC IV, and therefore the next steps consist of extending BPMN with concepts for structured data types and providing a translation into an executable agent language. We therefore plan to combine BPMN with OWL [16] and OWL-S in order to obtain knowledge representation together with reasoning mechanisms. Another interesting research task concerns the mapping into the opposite direction. Since multi-agent systems are naturally complex, a good visualisation of their workflow would increase the understanding of existing systems significantly.

References

1. Group, O.M.: Business process modeling notation (BPMN) specification. Final Adopted Specification dtc/06-02-01, OMG (2006) http://www.bpmn.org/Documents/OMGFinalAdoptedBPMN1-0Spec06-02-01.pdf
2. Ouyang, C., van der Aalst, W., Dumas, M., ter Hofstede, A.: Translating BPMN to BPEL. Technical Report BPM-06-02, BPMCenter.org (2006)
3. Desrocher, A.A., Deal, T.J., Fanti, M.P.: Complex-valued token petri nets. In: IEEE transactions on automation science and engineering (IEEE trans. autom. sci. eng.). Volume vol. 2., Institute of Electrical and Electronics Engineers, Piscataway, NJ, ETATS-UNIS (2004) (Revue) (2005) pp. 309–318

4. Murata, T.: Petri nets: Properties, analysis and applications. Proceedings of the IEEE **77**(4) (1989) 541–580
5. Vidal, J.M., Buhler, P., Stahl, C.: Multiagent systems with workflows. IEEE Internet Computing **8**(1) (2004) 76–82
6. Committee, O.: Web services business process execution language (WS-BPEL) version 2.0. Technical report, Oasis (2005)
7. Aalst, W.: The Application of Petri Nets to Workflow Management. The Journal of Circuits, Systems and Computers **8**(1) (1998) 21–66
8. Moldt, D., Wienberg, F.: Multi-agent-systems based on coloured petri nets. In: ICATPN. (1997) 82–101
9. Koehler, J., Hauser, R.: Untangling unstructured cyclic flows — a solution based on continuations. In Meesman, R., Tari, Z., eds.: CoopIS/DOA/ODBASE 2004. Volume 3290 of LNCS., Springer-Verlag (2004) 121–138
10. Casella, G., Mascardi, V.: From AUML to WS-BPEL. Technical Report DISI-TR-06-01, Dipartimento di Informatica e Scienze dell'Informatione, Università di Genova (2006)
11. Mantell, K.: From UML to BPEL — model driven architecture in a web services world. Technical report, IBM (2005) http://www-128.ibm.com/developerworks/webservices/library/ws-uml2bpel/.
12. Bauer, B., Müller, J.P., Odell, J.: Agent UML: A formalism for specifying multi-agent software systems. In Ciancarini, P., Wooldridge, M., eds.: Agent-Oriented Software Engineering, 1st International Workshop, AOSE 2000, Revised Papers. Volume 1957 of LNCS., Springer-Verlag (2001) 91–104
13. Bozzo, L., Mascardi, V., Ancona, D., Busetta, P.: COOWS: Adaptive BDI agents meet service-oriented computing – extended abstract. In Gleizes, M.P., Kaminka, G.A., Nowé, A., Ossowski, S., Tuyls, K., Verbeeck, K., eds.: Proceedings of the 3rd European Workshop on Multi-Agent Systems (EUMAS'05), Koninklijke Vlaamse Academie van Belie voor Wetenschappen en Kunsten (2005) 473–484
14. Rao, A.S.: AgentSpeak(L): BDI agents speak out in a logical computable language. In van Hoe, R., ed.: Agents Breaking Away, 7th European Workshop on Modelling Autonomous Agents in a Multi-Agent World, MAAMAW'96. Volume 1038 of LNCS., Eindhoven, The Netherlands, Springer Verlag (1996) 42–55
15. Walton, C.: Uniting agents and web services. In: Agentlink News. Volume 18. AgentLink (2005) 26–28
16. McGuinness, D.L., van Harmelen, F.: Owl web ontology language. w3c recommendation (2004) http://www.w3.org/TR/owl-features/.

Associated Topic Extraction
for Consumer Generated Media Analysis

Shinichi Nagano, Masumi Inaba, Yumiko Mizoguchi, and Takahiro Kawamura

Corporate R&D Center, Toshiba Corporation, Japan
1, Komukai-Toshiba-cho, Saiwai-ku, Kawasaki-shi, 212-8582, Japan
`shinichi3.nagano@toshiba.co.jp`

Abstract. This paper proposes a new algorithm of associated topic extraction, which detects related topics in a collection of blog entries referring to a specified topic. It is a partial feature of our product reputation information retrieval service whose aim is to detect product names rather than general terms. The main feature of the algorithm is to evaluate how important a topic is to the collection, according to the popularity of blog entries through Trackbacks and comments. Another feature is to utilize product ontology for topic filtering, which extracts products relevant to or similar to a specified product. The paper also presents a brief evaluation of the algorithm, in comparison with TF-IDF. In respect to the evaluation, it can be concluded that the proposed algorithm can capture users' impressions of associated topics more accurately than TF-IDF.

1   Introduction

Consumer Generated Media (CGM) have been one of major "word-of-mouth" media both for consumers and companies. The innovation of Web technologies produces a variety of Web services, which make it easy for consumers to create information or publish their opinions for themselves in the Internet. Blog is a typical Web service. Some consumers write reviews of products or describe their impressions or experiences of them, and others compare a specific product with similar products or with products competitive with it. The blogs often furnish blog readers with information that influences their purchase decisions. Moreover, the readers who buy the product may write blog entries commenting on it. Thus, this "word-of-mouth" cycle forms the wisdom of crowds and has the potential to grow large.

CGM has a strong impact not just on purchase decision processes by customers and but on product marketing strategies by companies. Analysis of CGM has actually been a challenging issue in product marketing. One of basic analysis methods is a topic word ranking, which is a statistical analysis of keywords used in search engines. An advanced method computes the frequency of occurrence of characteristic words appearing in blog contents. These help to discover either the demands on products or interests of consumers. Another method is a reputation analysis, which extracts reputation expressions from blog contents commenting on a specified product to summarize its reputation in terms of both quantity

J. Huang et al. (Eds.): SOCASE 2007, LNCS 4504, pp. 107–117, 2007.

and quality. It enables comparison of the product with its competing products in topicality and popularity, leading to understanding of the product positioning in the market. These methods are in early phase of application to practical product marketing. Some companies begin to provide reputation analysis features as public Web services, and others start the conduct of consumer marketing research based on reputation analysis.

We have developed a reputation analysis system[1], which retrieves blog entries commenting on a specified product, and then extracts reputation expressions and related products from the blog entries. The main feature of the system is that it analyzes the contents of retrieved blogs using ontology. This makes it possible (1) to indicate the overall rating of the product reputation (positive vs. negative), (2) to extract associated products that are much discussed in the blog entries, and (3) to sort the blog entries by reputation relevance and blog popularity.

This paper refers only to associated topic extraction in the service and proposes an extraction algorithm. The purpose of associated topic extraction is to suggest products similar to or relevant to a specified product. We propose a new algorithm for associated topic extraction. The method first extracts associated products through morphologic analysis of a collection of blog entries commenting on the specified product, and then computes the degree of their relevance. The main feature is the evaluation of how important a topic is to the collection, according to the popularity of blog entries, through Trackbacks and comments. Another feature is the utilization of product ontology for topic filtering to extract products relevant to or similar to the given product. The paper also briefly evaluates the proposed method in comparison with TF-IDF, a widely-used topic detection algorithm.

The reminder of this paper is organized as follows. First, Section 2 presents an overview of our reputation analysis system. Second, Section 3 addresses the issue of associated topic extraction and then Section 4 illustrates the proposed method. Next, Section 5 evaluates the algorithm, and finally, Section 6 concludes the paper.

2   Our Reputation Analysis System

2.1   System Overview

Our reputation analysis system is a semantic-based information retrieval of product reputations from blogs. It first retrieves blog entries commenting on a specified product, extracts reputation expressions and similar products from the retrieved blog entries, and then summarizes the reputation information on the target product. The main feature of the system is the semantic analysis of sentences in the retrieved blog entries using ontology, which is the description of concepts and their relationships. The ontology enables the following three features: (1) p/n (positive/negative) determination of the product reputation, (2) associated product extraction, and (3) blog sorting and filtering by reputation relevance. Each of the features are briefly described in the next subsection.

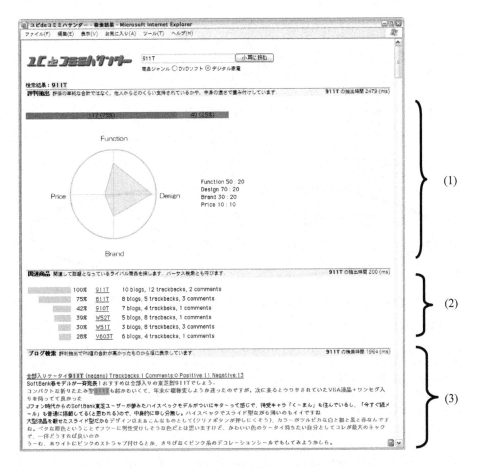

Fig. 1. A result of product reputation analysis

A result of product reputation analysis for a cellular phone is shown in Fig.2.1, and the architecture of our system is illustrated in Fig.2.

2.2 Three Features

P/n determination: The p/n determination is one of the text summarization technologies. It performs the morphologic analysis and the syntactic analysis of blog contents retrieved from the Internet, and then evaluates a positive or negative rating of the blog contents. A feature is that a rating of each blog entity is biased using link structures of Trackbacks and comments specific to blogs. One of the bias rules is to emphasize blogs written by an alpha blogger, who is regarded as an opinion leader respected in blog communities. Another feature is that a p/n rating of each sentence is biased using Japanese ontology that includes more than a hundred thousand concepts necessary to analyze reputation expression. An is-a relation in ontology determines strength of sensitive expressions, a

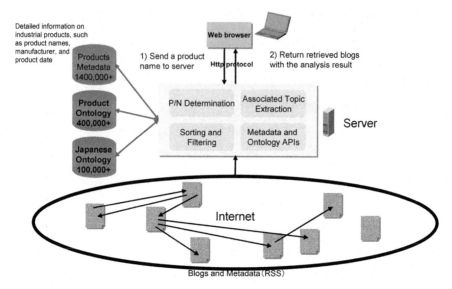

Fig. 2. Architecture of product reputation analysis system

part-of relation enables indirect rating of insensitive expressions, and an instance-of relation uniformalizes different expressions attributed to the same concept.

Fig.2.1(1) illustrates the result of the p/n determination, where two bars represent the positive rating (75%) and negative rating (25%) in total, and a radar chart shows the summation of positive and nagative ratings for each of four predefined viewpoints, function, design, brand, and price.

Associated topic extraction: The associated topic extraction is one of the text mining technologies based on term frequency and document frequency. The purpose is to suggest similar or comparable products that are much discussed in blogs referring to the given product. The function extracts product names by the morphologic analysis and the product ontology filtering, and then evaluates how important each extracted product is to a collection of the retrieved blog entries. A feature is that a rating is biased using correlation between blog entries. Another feature is that products irrelevant to the given product are excluded by referring to product ontology. This is another ontology, which represents products and hierarchical categorization by concepts and their relations, respectively, and includes more than 400 thousand products.

Fig.2.1(2) illustrates the result of the associated topic extraction, where six cellular phones are extracted with their normalized scores. Each of scores means how important the corresponding product is to a collection of retrieved blog entries.

Blog sorting and filtering: The sorting emphasizes blog entries receiving a lot of approvals through Trackbacks and comments, and blog entries containing

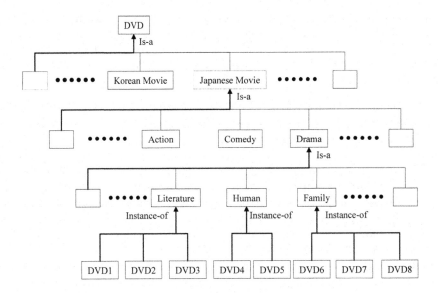

Fig. 3. Product ontology

a lot of positive or negative sentences. The filtering excludes promotional and marketing blogs, and blogs that just quote other blogs or articles and contain less impressions and reviews written by the bloggers themselves.

Fig2.1(3) shows selected useful blog entries, where positive and negative sentences, the specified product, and the associated products are highlighted.

2.3 Ontology and Metadata

We have developed three databases, product metadata, product ontology, and Japanese ontology. First, product metadata contains data describing products, such as product names, manufacturers, specifications, prices. It contains more than one million products, each of which is described in a form of RDF. Second, product ontology contains more than 400 thousand products including DVDs, books, and home appliances. It is composed of an is-a relation representing hierarchical categorization of products and an instance-of relation representing mappings of individual products with product categories. Third, Japanese ontology contains sensitive expressions necessary for reputation analysis of blog sentences. Both ontologies are described in a form of OWL. A fragment of the product ontology is illustrated in Fig.3.

3 Associated Topic Extraction

The purpose of associated topic extraction is to suggest similar or comparable products that are much discussed in retrieved blog entries commenting on a specified product. The straightforward approach to the extraction is to count

frequency of occurrence of proper nouns in blog entries through the morphologic analysis, where any blog entry and any noun appearing in blog entries
are uniformly enumerated. Although it may appear to be a plausible approach
such a uniform approach does not accord with blog readers' impressions for the
following reasons.

1. A reputation blog reader has a considerable impression corresponding to the
   number of Trackbacks and comments the blog entry received, compared with
   a reader of general web documents.
2. Since valuable information can often be found in blog threads composed of
   blog entries admired among blog communities, and thus such blog threads
   leave a significant impression on a reputation blog reader's mind.
3. A viewless blog entry should be eliminated, because the blog entry that
   express no opinion on products is not useful for a reputation blog reader. A
   typical blog is an advertisement blog that just provides product names with
   their specifications and prices.
4. The straightforward approach uselessly extracts a plurality of proper nouns
   other than product names. Such nouns are noisy for a reputation blog reader,
   and thus similar or comparable products mentioned in blog entries should
   be suggested.

4 Proposed Algorithm

4.1 Basic Idea

A main feature of the proposed extraction algorithm is that it finds a blog
thread, which is a set of blog entries spanned by Trackbacks, and then biases the
frequency occurrence of proper nouns according to the thread size. This is derived
from the following belief. Suppose that blog entries conversing about a target
product receive many Trackback pings. Then, the entries could be especially
interesting to many bloggers and products mentioned in the entries could be
similar to the target product or competitive with it.

Trackback is one of the main features in blog systems. It is a kind of references associating blog entries. A primary difference between blogs and other
Web documents is the bidirectional nature of blogs. Trackback is simply an ac-
knowledgement, which is a network signal (ping) sent from a originating blog
entry to a receiving blog entry. The receptor often publishes a link back to the
originator to indicate its reception. Another difference is a communication facilitating feature of Trackback. If a blogger writes a new entry commenting on an
entry found at another blog, then the commenting blogger can notify the other
blog with a "Trackback ping"; the receiving blog will typically display links to
all the commenting entries below the original entry. This enables conversations
spanning several blogs that blog readers can easily follow.

Another feature of the algorithm is that it utilizes product ontology. Product
ontology is composed of an is-a relation representing hierarchical categorization of products and an instance-of relation representing mappings of individual

products with product categories. Product ontology is used to find only similar or competitive products mentioned in the blog threads conversing on a target product, filtering out proper nouns other than product names and biasing products closer to the target product by referring to the distances of their categories on product ontology.

4.2   Proposed Algorithm

Given a set of blog entries conversing on a certain topic, the proposed algorithm extracts a list of similar, related, or comparable topics, which are much discussed in the given blog entries. The algorithm also provides the degree of association $AT(t)$ for each extracted topic t.

Let $TF(t)$ and $IF(t)$ be the topic occurrence frequency t and the inverse document frequency for topic t. Given a set of blog entries conversing on a certain product, then $AT(t)$ is defined as follows:

$$AT(t) = (TF_R(t) + TF_{IR}(t)) \times IF(t) \tag{1}$$

Note that our $TF_R(t)$, $TF_{IR}(t)$, and $IF(t)$ measures employing three bias rules are a refined version of the original TF-IDF measure. $TF_R(t)$ and $TF_{IR}(t)$ are computed by two bias rules utilizing correlation between blog entries, which are defined in subsections 4.3 and 4.4. $IF(t)$ is computed by the bias rule utilizing document frequency, which is defined in subsection 4.5.

4.3   Blog-Reference-Based Bias Rule

This rule is to bias blog entries according to the amount of commenting blog entries. It is derived from the belief that the bloggers could tend to appreciate blog entries according to the amount of Trackbacks and comments. Thus, blog entries receiving a lot of Trackback pings could raise the value of a Trackback ping.

The rule is defined by multiplication of a topic frequency $TF(t,i)$ and its biasing function $BR_R(t,i)$ for topic t and blog entry i.

$$TF_R(t) = \sum_i TF(t,i) \times BR_R(t,i) \tag{2}$$

$TF(t,i)$ is slightly different from the usually used definition, because blog entries are absolutely shorter than general text documents. Let $TP(i)$ be the number of topics which appear in blog entry i, and then $TF(t,i)$ is defined as follows.

$$TF(t,i) = \begin{cases} 1/TP(i) \\ \quad \text{if topic } t \text{ appears in blog entry } i \\ 0 \quad \text{otherwise} \end{cases} \tag{3}$$

Next, suppose that $R(i)$ denotes a set of blog entries referring to the receptor of a Trackback ping from blog entry i and that TB stands for the total number of

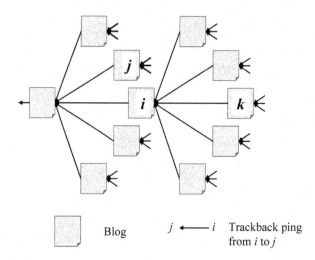

Fig. 4. Blog-reference-based and blog-referred-based bias rules

Trackbacks received by any blog entry in a given collection. Then, the reference-based bias function $BR_R(t, i)$ is defined as follows.

$$BR_R(t, i) = \sum_{j \in R(i)} \frac{TB(j) + 1}{TB} \tag{4}$$

$BR_R(t, i)$ biases commenting blog entry j according to the number of Trackback receptions by blog entry j.

4.4   Blog-Referred-Based Bias Rule

This rule is to bias blog entries according to the amount of commenting blog entries, in the same manner as the rule defined in subsection 4.3. The difference between the two rules are described below. Let i, j, k be blog entries, where i is commenting j and k is commenting i. This referred-based rule computes the value of i based on the value of k although the reference-based rule defined in subsection 4.3 computes the value of i based on the value of j.

The referred-based rule is defined by multiplication of a topic frequency $TF(t, i)$ and its biasing function $BR_{IR}(t, i)$ for topic t and blog entry i. $TF(t, i)$ is the same as Eq. 3.

$$TF_{IR}(t) = \sum_i TF(t, i) \times BR_{IR}(t, i) \tag{5}$$

Next, suppose that $IR(i)$ denotes a set of blog entries by which blog entry i is referred, Then, the referred-based bias function $BR_{IR}(t, i)$ is defined as follows.

$$BR_{IR}(t, i) = \sum_{k \in IR(i)} \frac{TB(k) + 1}{TB} \tag{6}$$

$BR_{IR}(t, i)$ biases commenting blog entry k according to the number of Trackback receptions by blog entry k.

4.5   Product-Occurrence-Based Bias Rule

This rule is a refined version of the inverse document frequency scheme, and computes how important a topic is to blog entries. The difference with the original measure is to bias the weight of the importance according to the document frequency. This is derived from the following belief; The document frequency of a topic would be a good measure of popularity in blog entries, but too high frequency would mean to be valueless because many bloggers know the topic. This bias rule is defined as $IF(t)$, where N and $DF(t)$ denote the number of blog entries in collection and the document frequency of topic t to the blog entries, respectively.

$$IF(t) = DF(t) \times \log(\frac{N}{DF(t)}) \tag{7}$$

4.6   Product Ontology Filtering

This filtering provides two features. The first feature is to filter out proper nouns other than product names, given a collection of proper nouns extracted by the morphologic analysis of blog entries. Another feature is a category filtering based on class distance. This excludes products unrelated to the target products. For example, a refrigerator categorized in home appliances is excluded from similar or competing products for a novel categorized in books.

5   Evaluation

We evaluate how correctly the proposed algorithm captures users' impression on associated products.

Evaluation is done as follows. We have first invited twenty public users to manually extract associated products from each of three real blog threads commenting on DVDs. The blog threads are composed of more than 10 blog entries, and structured through Trackback links and comments. The answers are used as correct data for evaluation. Next, we apply both the proposed algorithm and TF-IDF to the same data set of blog entries, and then compare scores of associated topics. The evaluation measure is how closer computed scores are to the correct answers.

Table 1 shows comparison of the proposed algorithm and TF-IDF with the correct answers. The first row represents product names, where ten DVDs are extracted as associated topics from blog entries commenting on a target DVD. The second row represents the answers, which are summation of users' answers for each associated products. Note that the data are normalized for DVD1. The third and forth row show the scores computed by the proposed algorithm and TF-IDF, respectively. Both data are also normalized for DVD1.

Table 2 averages the difference of the computed score with the answer for each DVD. It shows that the difference of the proposed algorithm is closer than

Table 1. Score comparison of associated topics

Products	Answers	Proposed method	TF-IDF
DVD1	100%	100%	100%
DVD2	50%	77.8%	29.5%
DVD3	39%	77.8%	29.5%
DVD4	39%	36.8%	15.4%
DVD5	22%	8.3%	4.2%
DVD6	22%	8.3%	4.2%
DVD7	17%	8.3%	4.2%
DVD8	6%	8.3%	4.2%
DVD9	33%	49.4%	29.6%
DVD10	11%	19.4%	14.1%

Table 2. Differences with the correct answer

	Difference
$|$Answer − Proposed$|$	7.2%
$|$Answer − TFIDF$|$	12.2%

TF-IDF. Thus, in respect to this evaluation, we can conclude that the proposed algorithm captures users' impression more accurately than TF-IDF.

6   Conclusions

We have proposed a new algorithm which finds associated topics in a collection of blog entries referring to a given topic. The evaluation result shows that the proposed algorithm can capture users' impressions on associated topics more accurately than TF-IDF. We are planning to evaluate precision and performance measures of the proposed algorithm in details. We are also going to apply our CGM analysis system to practical product marketing.

References

1. Kawamura, T., Nagano, S., Inaba, M., Hasegawa, T., Ohsuga, A.: Ubiquitous Metadata Scouter - ontology brings blogs outside. In: Proceedings of 1st Asian Semantic Web Conference (ASWC2006). (2006)
2. Kleinberg, J.: Bursty and hierarchical structure in streams. In: Proc. of the 8th ACM SIGKDD International Conference on Knowledge Discovery and Data Mining. (2002) 1–25
3. Kleinberg, J.: Authoritative sources in hyperlinked environment. Journal of the ACM **46** (1999)
4. Chakrabarti, S.: Mining the web. (2003)
5. Fujimura, K., Inoue, T., Sugizaki, M.: The EigenRumor algorithm for ranking blogs. In: Proceedings of the WWW2005 Workshop on the Weblogging Ecosystem. (2005)

6. Kamvar, S., Schlosser, M., Garcia-Molina, H.: The EigenTrust algorithm for reputation management in P2P networks. In: Proceedings of 12th International World Wide Web Conference. (2003)
7. Mishne, G., de Rijke, M.: A study of blog search. In: Proceedings of 28th European Conference on Information Retrieval (ECIR). (2006)
8. Fujiki, T., Nanno, T., Suzuki, Y., Okumura, M.: Identification of bursts in a document stream. In: Proceedings of First International Workshop on Knowledge Discovery in Data Streams. (2004)
9. Nanno, T., Fujiki, T., Suzuki, Y., Okumura, M.: Automatically collecting, monitoring, and mining japanese weblogs. In: Proceedings of 13th International World Wide Web Conference. (2004)
10. Facca, F., Lanzi, P.: Mining interesting knowledge from weblogs: a survey. Data and Knowledge Engineering **53** (2005) 225–241
11. Allan, J., ed.: Topic Detection and Tracking: Event-based Information Organization. Kluwer Academic Publishers (2002)
12. Benjamin, Trott, M.: Trackback technical specification. http://www.movabletype.org/docs/mttrackback.html (2002)

An MAS Infrastructure for Implementing SWSA Based Semantic Services

Önder Gürcan[1], Geylani Kardas[2], Özgür Gümüs[1],
Erdem Eser Ekinci[1], and Oguz Dikenelli[1]

[1] Ege University, Department of Computer Engineering,
35100 Bornova, Izmir, Turkey
{onder.gurcan,ozgur.gumus,oguz.dikenelli}@ege.edu.tr
erdemeserekinci@gmail.com
[2] Ege University, International Computer Institute,
35100 Bornova, Izmir
geylani.kardas@ege.edu.tr

Abstract. The Semantic Web Services Initiative Architecture (SWSA) describes the overall process of semantic service execution in three phases: discovery, engagement and enactment. To accomplish the specified requirements of these phases, it defines a conceptual model which is based on semantic service agents that provide and consume semantic web services and includes architectural and protocol abstractions. In this paper, an MAS infrastructure is defined which fulfills fundamental requirements of SWSA's conceptual model including all its sub-processes. Based on this infrastructure, requirements of a planner module is identified and has been implemented. The developed planner has the capability of executing plans consisting of special tasks for semantic service agents in a way that is described in SWSA. These special tasks are predefined to accomplish the requirements of SWSA's sub-processes and they can be reused in real plans of semantic service agents both as is and as specialized according to domain requirements.

1   Introduction

The Semantic Web Services Initiative Architecture (SWSA) committee, which has been contributed by the Semantic Markup for Services (OWL-S)[1], Web Service Modeling Ontology (WSMO)[2] and Managing End-to-End Operations-Semantics (METEOR-S)[3] working groups, has created a set of architectural and protocol abstractions that serve as a foundation for Semantic Web service technologies[1]. The proposed SWSA framework builds on the W3C Web

---

[1] The OWL Services Coalition: Semantic Markup for Web Services (OWL-S), 2004, http://www.daml.org/services/owl-s/1.1/

[2] Web Service Modeling Ontology (WSMO) Working Group, http://www.wsmo.org/

[3] Managing End-to-End Operations-Semantics (METEOR-S) Working Group, http://lsdis.cs.uga.edu/projects/meteor-s/

J. Huang et al. (Eds.): SOCASE 2007, LNCS 4504, pp. 118–131, 2007.

Services Architecture working group recommendation[4] and attempts to address all requirements of semantic service agents: dynamic service discovery, service engagement, service process enactment and management, community support services, and quality of service (QoS). This architecture is based on the multi-agent system (MAS) infrastructure because the specified requirements can be accomplished with asynchronous interactions based on predefined protocols and using goal oriented software agents.

The SWSA framework describes the overall process of discovering and inter-acting with a Semantic Web service in three consecutive phases: (1) candidate service discovery, (2) service engagement, (3) service enactment. The SWSA framework also determines the actors of each phase, functional requirements of each phase and the required architectural elements to accomplish these re-quirements in terms of abstract protocols. Although it defines a detailed concep-tual model based on MAS infrastructure and semantic web standards, it does not define the software architecture to realize this conceptual model and does not include the theoretical and implementation details of the required software architecture.

There have been a few partial implementations to integrate web services and FIPA compliant agent platforms. WSDL2Jade [2] can generate agent ontologies and agent codes from a WSDL input file to create a wrapper agent that can use external web services. WSDL2Agent [3] describes an agent based method for migrating web services to the semantic web service environment by deriving the skeletons of the elements of the Web Service Modeling Framework (WSMF) [4] from a WSDL input file with human interaction. WSIG (Web Services In-tegration Gateway) [5] supports bi-directional integration of web services and Jade agents. WS2JADE [6] allows deployment of web services as Jade agents' services at run time to make web services visible to FIPA-compliant agents through proxy agents. But these tools only deal with the integration of agents and external web services and do not provide any mechanism to realize the entire architectural and protocol abstractions addressed by the SWSA framework. It's clear that there must be environments which will simplify the development of SWSA based software systems for ordinary developers.

The main contribution of this paper is to define a software platform which fulfills fundamental requirements of SWSA's conceptual model including all its sub-processes. Then, these sub-processes are modeled as reusable plans for de-velopment of semantic service agents. And the second contribution of this paper is to develop a planner that has the capability of executing these kinds of plans. So, the developed planner has the innovative features listed below:

- Definition of reusable template plans that includes abstract task structures for SWSA's sub-processes and usage of these templates for generating agents' real plans by specializing these abstract tasks
- Support for recursion on the plan structure

---

[4] W3C Web Services Architecture Working Group, Web Services Architecture Rec-ommendation, 11 February 2004, http://www.w3.org/TR/ws-arch/

– Constitution of composite services by using reusable semantic service agent plans

The paper is organized as follows: in section 2, the proposed architecture of the Software Platform for the SWSA framework is given. Planning requirements of SWSA is discussed in section 3. Section 4 introduces the planner component of the platform. Conclusion and future work are given in section 5.

2   Semantic Service Platform Architecture

It is apparent that SWSA describes the architecture extensively in a conceptual base. However it doesn't define required details and theoretical infrastructure to realize the architecture. Hence, we propose a new software platform in which above mentioned fundamental requirements of all SWSA's sub-processes (service discovery, engagement and enactment) are concretely fulfilled.

The software architecture of the proposed Semantic Service Platform is given in Figure 1. The platform is composed of two main modules called *Semantic Service Kernel* and *External Service Agent*.

The Semantic Service Kernel includes the required infrastructure and architectural components for an agent to execute sub-processes of SWSA. The agent's actions, to be used in semantic service discovery, engagement and enactment, are

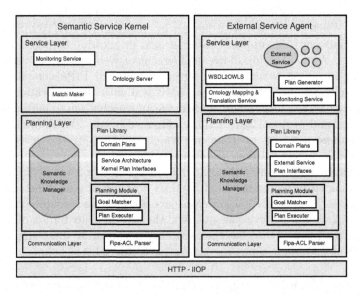

Fig. 1. The software architecture of the Semantic Service Platform: Two main models are Semantic Service Kernel and External Service Agent. The kernel provides architectural components for an agent to execute sub-processes of SWSA. On the other hand, External Service Agent provides integration of external services (either purely defined in WSDL or OWL-S) into the agent platform. Communication takes place via the well-known HTTP - IIOP (Internet Inter-ORB Protocol).

modeled as reusable plans and will be executed in a composite fashion by a planner. The sub-tasks, which compose the plan, will execute SWSA's sub-processes by invoking the related services. Invocation will be realized via predefined execution protocols.

The External Service Agent converts either WSDL or OWL-S defined external services into agents that are able to execute SWSA's defined processes. This module includes inner services like WSDL to OWL-S Converter, Ontology Mapper and Translator -that provides mapping of services into the platform's ontologies, stores those mapping ontologies and serves ontology translation- and Monitor service that monitors quality of service parameters. Planner component of the External Service Agent realizes registration of the related service into the platform and executes interaction plans concerning service engagement and enactment. Those plans are formed automatically during the creation phase of the External Service Agent and stored in the plan library as domain plans.

In the following subsections, we discuss details of how our proposed semantic service platform meets the requirements of the SWSA taking into consideration predefined SWSA sub-processes.

2.1 Realization of Service Discovery Process

In order to realize semantic service discovery, the platform services should be registered to a matchmaker and service clients should query on this matchmaker and have ability to interpret resultant service advertisements.

The Matchmaker service of the Semantic Service Platform stores capability advertisements of registered services as OWL-S profiles. As previously implemented in SEAGENT environment [7], the capability matching of the requested and registered service advertisements herein, is also based on the algorithm given in [8] and deals with semantic distances between input/output parameter concepts of related services. The details of the implemented capability matching may be found in [9] and [10]. However, in addition to the above mentioned matching algorithm, Matchmaker service of our proposed platform also supports semantic match only on types of services (excluding input/output parameter match). Therefore, a client may indicate his/her preferred capability matching approach to the matchmaker and matchmaker performs capability matching upon this client's preference.

Based on domain knowledge of the related application, the Semantic Service Platform provides a *meta-profile* definition for platform services those to be registered, discovered and invoked within the platform. Hence, in this approach, Semantic Service Kernel plans, that include client invocation codes, may be prepared easily by only using those predefined meta-profiles of services. However this naturally exposes an ontology mapping requirement when an outer service, that is needed to be included in the platform, has a different profile model than platform's meta-definitions. It is aimed to bring a solution to this problem by using capabilities of the ontology mapping and translation service of the External Service Agent. So, advertisement plan of the External Service Agent supports platform administrators to be able to map platform's meta-profile with the related external service's ontology by using mapping service via a user interface.

It should be noted that discovery process of the client, has already been defined as a reusable plan template in the Semantic Service Kernel. So, the content of this plan template is determined during domain based application development and this creates the application dependent plan of the discovery process. The client agent, which uses the related created discovery plan, first sends the required service's profile to the matchmaker service and receives advertisement profiles of semantically appropriate services. The suitable communication protocol and content language for the client has already been designed and implemented for OWL-S services [9].

Another important task of the discovery process is the service selection policy of the requester client. The Matchmaker of the platform may return a collection of suitable service profiles for the client's requests in many cases and client should apply a policy into the result collection to select the service(s) for further engagement and enactment processes. The Semantic Service Platform provides extensible service selection policy structures for plan designers to add various selection criteria into the service user agent plans.

2.2 Realization of Service Engagement Process

After completion of the service selection, the client-service engagement process begins. The engagement process has 2 stages: (1) negotiation on quality of service metrics between client and service agents and (2) agreement settlement.

The first stage of the engagement process includes determination of the exact service according to quality of service (QoS) metrics. Currently, there exists no standard for the service quality metrics. However, during the exact service determination, our proposed service platform utilizes some QoS parameters (like service cost, run-time, location, etc.) defined in various studies [11,12] which address this issue. When both sides (client and service) agree on the quality metrics, the first stage of the process is finished.

The engagement process is completed after determined service's OWL-S process ontology and QoS parameters are sent to the Monitoring Service for being monitored during service execution.

2.3 Realization of Service Enactment Process

Conceptually, enactment can be defined as the invocation of the engaged service by the client agent. However, in fact, enactment includes more than just invocation and it should take into consideration of monitoring, certification, trust and security requirements of service calls.

Execution of composite semantic services (modeled by using OWL-S) is maintained in the platform by means of a planning approach. The approach herein, provides definition of service templates for each atomic service of the composite service and realizes composition of the service by linking those atomic sub-processes.

Service execution also requires monitoring of the invocation according to the engagement between client agent and the server. Monitoring services of the Semantic Service Kernel and External Service Agent both monitor execution of services and control whether current interaction conforms into the predefined QoS metrics and engagement protocol or not. Hence, the Monitoring service of the External Service Agent informs the platform's monitoring service about the produced values of the quality metrics during service execution. According to the state of the ongoing interaction, the informed client agent may change his task execution behavior as defined in his enactment plan.

3   Planning Requirements of SWSA

The Semantic Service Kernel includes the required infrastructure and architectural components for an agent to execute subprocesses of SWSA. Such an environment simplifies the overall process of executing semantic web services for ordinary developers. Client agent(s) in this environment, must provide plans to be used in semantic service discovery, engagement and enactment. Hence, in our platform, we modeled these plans as reusable plans that are defined using the well known Hierarchical Task Network (HTN) planning structures. HTN Planning is an AI planning methodology that creates plans by task decomposition. This decomposition process continues until the planning system finds primitive tasks that can be performed directly. The basic idea of HTN planning was in the mid-70s [13,14], and the formal underpinnings were developed in the mid-90s [15]. In an HTN planning system, the objective is to accomplish a partially-ordered set of given tasks (plan) and a plan is correct if it is executable, and it accomplishes the given tasks. That is, the main focus of an HTN planner is to perform tasks, while a traditional planner focuses on achieving a desired state.

The planner of our MAS development framework, called SEAGENT, is based on the HTN planning framework presented by Sycara et. al [16] and DECAF architecture [17]. In SEAGENT, tasks might be either complex (called behaviours) or primitive (called actions). Each plan consists of a root task (behaviour) which is a complex task itself consisting of sub-tasks to achieve a predefined goal. Behaviours hold a 'reduction schema' knowledge that defines the decomposition of the complex task to the sub-tasks and the information flow between these sub-tasks and their parent task. The information flow mechanism is as follows: each task represents its information need by a set of *provisions* and the execution of a task produces *outcomes*, and there are links that represents the information flows between tasks using these provision and outcome slots. Actions, on the other hand, are primitive tasks that can be executed directly by the planner. Also, each task produces an outcome state after its execution. Default outcome state is "OK" and usually it is not shown. This outcome state is used to route the information flow between tasks.

Figure 2 shows the task tree of the HTN plan for service execution of client agents. "Execute Service" task represents the service execution process which was

proposed by SWSA and is the root task of the plan that needs abstract charac-
terizations of candidate services in order to be executed. It is composed of three
sub-tasks: "Discover Candidate Services", "Engage with a Service" and "Enact
Service" (decomposition of these tasks are not shown in this figure). "Discover
Candidate Services" task inherits the abstract characterization of candidate services from its parent task and produces service profiles of candidate services after
its execution completes. Then these service profiles are passed to "Engage with
a Service" task via a provision link and then the execution of this task begins.
"Engage with a Service" task finishes by producing two outcomes: selected service provider and service agreement. These outcomes are consumed by "Enact
Service" task in order to complete the final part of the plan. This task executes
selected service and passes the output list of its execution to its parent task.

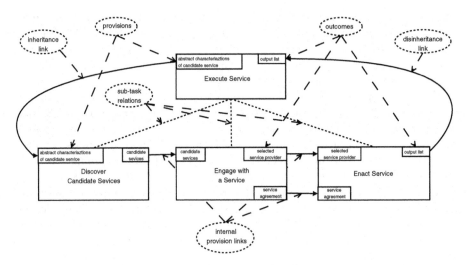

Fig. 2. HTN plan for SWSA based Semantic Service Execution

As discussed above, SWSA processes are controlled and monitored by Semantic Service Kernel. Planner is at the heart of the architecture which controls
the workflows of the SWSA processes. Each of these processes are modeled as
reusable plans and the developed planner must have the innovative features in
order to have the capability of executing these kind of plans. These features are
listed below:

- *Definition of reusable template plans that includes abstract task structures
  for SWSA's subprocesses and usage of these templates for generating agent's
  real plans by specializing these abstract tasks.*

Some parts of processes introduced by SWSA is abstract because, they
change according to domain. Hence, it must be possible to generate a
template plan for SWSA and realize it on specific domains. Such template plan must include variable (or abstract) tasks which can be specified based on the domain requirements. So, the plan structure must

provide mechanisms to define and specify such variable task constructs for developers. Without such a planning mechanism, it is impossible to define reusable plan templates for executing SWSA processes. For example, in service discovery, the client agent, first forms a query using the required service's profile, sends it to the matchmaker service, receives advertisement profiles of semantically appropriate services and finally applies a service selection policy to return a collection of suitable profiles. The reusable template plan for this service discovery process is illustrated in Figure 3. As shown in the figure, "Form a Query for Service Discovery" task and "Select Service(s)" task are abstract. Former is abstract because the query could be formed either according to service type or input/output parameter types or etc. Latter is abstract because developers should be able to use various service selection criteria.

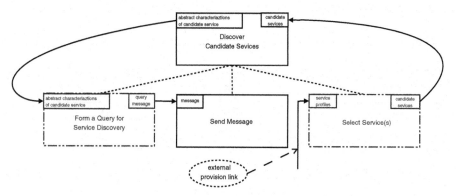

Fig. 3. Decomposed view of the task for service discovery

– *Support for recursion on the plan structure.*

By a recursion we mean a situation where one instance of a task is an ancestor in the planning tree of another instance of the same task. Consider the engagement process of the client agent given in section 2.2. At first, one of the candidate services are chosen and then completeness of all service invocation parameters is assured. After the assurance, negotiation on QoS metrics and agreement settlement are performed. If either assurance or negotiation tasks fail, the engagement process will be restarted for unselected services (Figure 4). To provide this iteration the plan structure must have support for recursion. Such support needs the ability for a task to contain itself as a sub-task (in Figure 4, "Engage with a Service" contains itself as a sub-task).

– *Constitution of composite services by using semantic service execution plans in agent's real plans.*

To generate domain dependent service execution plans, developer must realize "Execute Service" template plan shown in Figure 2. Consider the

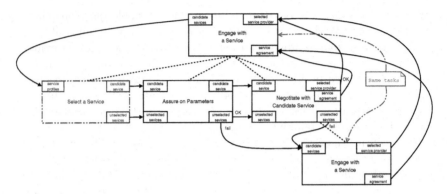

Fig. 4. Decomposed view of the task for service engagement

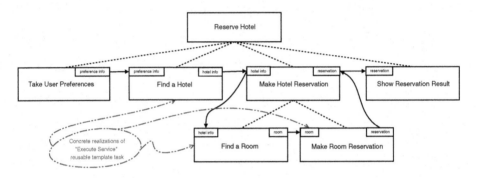

Fig. 5. Hotel reservation plan

HTN plan for reserving a hotel room shown in Figure 5. Reserving a hotel is as follows: first user preferences are taken, and according to these preferences a hotel is found, then it is tried to make reservation with that hotel and finally results are shown to the user. In this plan, "Find a Hotel", "Find a Room" and "Make Room Reservation" sub-tasks of the plan are concrete realizations of "Execute Service" task. They are connected with their provisions and outcome slots, and because they are domain dependent plans they know what input parameters they will take.

4   SEAGENT Planner

4.1   Internal Architecture

The overall structure of the planner architecture (Figure 6) is designed to execute HTN structure(s) that includes complex and primitive tasks. In order to execute a plan, the planner dynamically opens the complex root task using the 'reduction schema' knowledge, and this reduction continues until the planner finds actions (directly executable tasks), and then the planner executes these

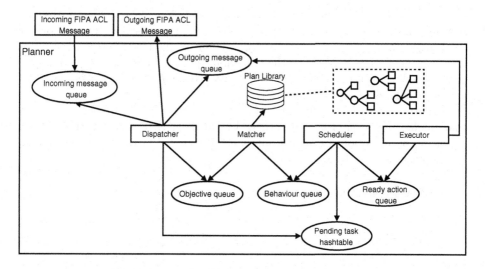

Fig. 6. Overall structure of SEAGENT planner

actions. Propagation of output values of the executed task to other dependent task(s) is handled by the "plan structure" itself. The planner is composed of four functional modules: dispatcher, matcher, scheduler and executor. Each module runs concurrently in a separate Java thread and uses the common data structures. All together, they match the goal extracted from the incoming FIPA-ACL message to an agent plan, schedule and execute the plan following the predefined HTN based plan. In the following, we briefly explain responsibilities of each module during plan execution.

The planner is responsible for processing incoming and outgoing messages, matching (finding), scheduling and executing tasks. When a FIPA-ACL message is received by the agent, it is enqueued to incoming message queue by the communication infrastructure layer. Dispatcher checks incoming message queue for the existence of an incoming message, if so, it parses the message and checks whether it is part of an ongoing conversation or not. If so, then the dispatcher finds out the task waiting for that message from pending task hash table, and sets the provision(s) of that task. If the incoming message is not part of an ongoing conversation, then the dispatcher creates a new objective, puts it to the objective queue.

Matcher is responsible for matching the incoming objective to a pre-defined plan by querying the plan library which is constructed using plan ontology and match ontology. When a new objective is occurred, matcher gets message information from it, and tries match a plan from the plan library for that objective. If match succeeds, Matcher creates an instance of the plan and enqueues it to behaviour queue.

Scheduler's role is to determine the execution time of each behaviour. Scheduler gets behaviours from the behaviour queue and prepares them for execution. If all of the provisions of the behaviour are set, Scheduler reducts that behaviour by calling its **reduct()** method - decomposes the behaviour to its sub-tasks. All sub-tasks are checked whether their all provisions are set or not. If provisions of a sub-task are set, then, if it is an action it is put to the ready action queue, else (if it is a behaviour) it is put to the behaviour queue. If not, the sub-task is put to the pending task hash table to wait for its provisions to be set. They may be action or behaviour.

Executor checks the ready action queue and if there are pending actions on that queue, it gets and executes them by invoking their **Do()** method. After execution of an action, produced outcomes are passed to the dependent task's provisions internally by the plan itself.

4.2   SEAGENT Plan Structure

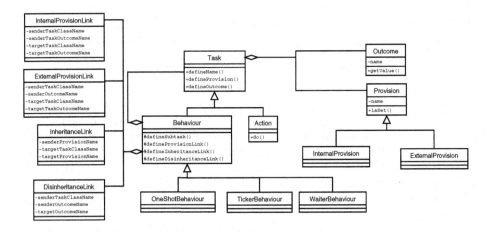

Fig. 7. Components of SEAGENT plan structure

Components of SEAGENT plan structure are shown in Figure 7 (only key attributes and operations are shown). Tasks are represented with **Task** class and have a name describing what it is supposed to do and have zero or more provisions and outcomes. Provisions might be of two types: internal provision and external provision. Internal provisions are provisions whose value's are internally set within the plan, in other words, value of internal provisions is determined by outcome of another tasks. External provisions, on the other hand, is set externally, with an incoming FIPA-ACL message. Incoming messages from another agents are routed to the external provisions of pending tasks. This routing is done according to *conversation-id* and *in-reply-to* fields of the incoming message. There are various types of behaviours, such as **OneShotBehaviour** which is executed only once, **TickerBehaviour** which is executed periodically and **WaiterBehaviour** which is executed after an amount of time is passed. All specifications about behaviours and actions is hold by themselves and, these specifications are described by using **defineXXX()** methods.

To write an action, it is enough to extend `Action` class, define provisions and outcomes using `defineProvision()` and `defineOutcome()` methods and finally implement `do()` method. Writing behaviours is a bit different because, behaviours have no `do()` method, and they only hold the specifications about its sub-tasks, and relationships between these sub-tasks. Also they may contain provision and outcomes, and, definition of provisions and outcomes is just as in actions. To define sub-tasks, `defineSubtask( String className )` method is used by giving the absolute class name of the task to be added to this behaviour. For example, "Enact Service" task in Figure 2 can be defined in "Execute Service" parent task as `defineSubtask( EnactService.class.getName() )`. And also provision links must be defined in order to satisfy information flow by using provision link definition methods[5].

4.3   SWSA Support in Planning Level

As mentioned in section 3, SEAGENT planner has innovative features to handle the workflows of the SWSA processes. This section shows how these features are satisfied by SEAGENT planner. The built-in support makes implementing overall processes of SWSA easier for developers.

Template Plan Support. In SEAGENT it is possible to create template (generic) plans. The basic idea resembles abstract class logic in object orientation. That is, to construct a generic plan, describe the main characteristics of a plan leaving some special pieces (tasks) unimplemented. So that in the future they may be specialized and used (remember the service discovery plan in Figure 3). Tasks defined in reusable template plan can be extended (by redefinition of abstract tasks) to concrete plans. The important point here is the confliction of the specifications of the abstract task to be extended (provision and outcome definitions) and specifications of the extended concrete task. To implement template plans, construct the plan as a normal plan but use abstract task definitions where you want to make abstract tasks.

In SEAGENT, to define sub-tasks, `defineSubtask()` method is used by giving the class name of the sub-task as a parameter. If we want to define an abstract sub-task, all we need to do is to give the name of the abstract sub-task indirectly. This can be done by using an abstract method that returns the name of the sub-task, and passing this abstract method as parameter to `defineSubtask()` method. For example, "Select Service(s)" abstract task in Figure 3 can be defined in "Discover Candidate Services" parent task by using `getSelectServicesTaskName()` abstract method (`defineSubtask( getSelectServicesTaskName() )`. Concrete realization of this plan is made by extending this plan and implementing the abstract methods that return the class names of the concrete tasks.

Recursion Support. Recursion is another capability of SEAGENT planner. By a recursion we mean a situation where one instance of a task is an ancestor

---

[5] `defineInternalProvisionLink()` and `defineExternalProvisionLink()`.

in the planning tree of another instance of the same task. This is usually used when we cannot satisfy something in a task and want to execute this task again but with different provisions. Since there is no restriction on defining sub-tasks, a task may contains itself as a sub-task. But the important point here is that the decomposition of the successor task must be in control, just like in recursive methods in traditional programming. In SEAGENT, the decomposition of a task is controlled via its provision's state, that is, if all provisions are set or there is no provision then the task decomposes. So, in a recursive HTN plan, recursion tasks must contain at least one provision and the value of this provision must be different from its ancestor's value (see Figure 4). Otherwise an infinite loop arises and our agent crashes.

Constitution of Composite Services Support. Constitution of composite services is simple in SEAGENT, because there are predefined reusable template plans for service execution (see Figure 2) in plan library and, developers can use these plans to construct their own domain dependent service execution plans and compose them just as building an ordinary plan (see Figure 5). The important point here is the satisfaction of input/output compatibility and this is easily handled by the correct concrete realizations of the abstract tasks.

5   Conclusion

The SWSA is currently a working group recommendation and describes abstract infrastructures and related processes for semantic web services and agents interaction in a conceptual base. We believe that this architecture brings a comprehensive model of software agents which utilize and provide semantic web services. However this architecture is a product of an initiative study and most of its components are only theoretically defined, not implemented. In this paper, a new MAS software platform, which aims to concretely fulfill fundamental requirements of the SWSA, has been introduced. We have modeled subprocesses of SWSA as reusable plans by HTN approach and provided a framework in which those plans can be executed in a composite fashion by agent planners. Hence, platform agents can accomplish execution of discovery, engagement and enactment processes for semantic web service interaction by employing those reusable and predefined HTN plans. We have also discussed necessary properties of an agent planner which can execute those defined plans. Such a planner has been implemented based on the SEAGENT platform.

In the paper we focused on the requirements of the planner for execution of SWSA subprocesses. But we are also working on the other parts of the software architecture. For example, service discovery mechanisms for the platform are fully operational. Semantic capability matching of services has already been implemented and platform agents are currently able to invoke semantic web services in proper to OWL-S standards.

Perhaps our major weakness, considering both the software and reusable agent plans, is seen in definition and design of the service engagement sub-process. QoS topics are currently being studied and they weren't addressed in detail by the service oriented computing community. Hence our QoS support during the engagement process is extremely primitive and only composes monitoring service. That support is also in its initial state. Security and trust mechanisms have not been considered yet in our implementation.

References

1. Burstein, M., Bussler, C., Zaremba, M., Finin, T., Huhns, M., Paolucci, M., Sheth, A., Williams, S.: A semantic web services architecture. IEEE Internet Computing **Volume 9 Issue 5** (2005) 72 – 81
2. Varga, L.Z., Ákos Hajnal, Werner, Z. In: Engineering Web Service Invocations from Agent Systems. Volume 2691. Lecture Notes in Computer Science (2003) 626 – 635
3. Varga, L.Z., Ákos Hajnal, Werner, Z. In: An Agent Based Approach for Migrating Web Services to Semantic Web Services. Volume 3192. Lecture Notes in Computer Science (2004) 381 – 390
4. Fensel, D., Bussler, C.: The web service modeling framework wsmf. Electronic Commerce Research and Applications **1** (2002) 113–137
5. Greenwood, D., Calisti, M.: Engineering web service - agent integration. In: SMC (2), IEEE (2004) 1918–1925
6. Nguyen, T.X., Kowalczyk, R.: Ws2jade: Integrating web service with jade agents. In: AAMAS'05 Workshop on Service-Oriented Computing and Agent-Based Engineering (SOCABE'2005). (2005)
7. Dikenelli, O., Erdur, R.C., Özgür Gümüs, Ekinci, E.E., Önder Gürcan, Kardas, G., Seylan, I., Tiryaki, A.M.: Seagent: A platform for developing semantic web based multi agent systems. In: AAMAS, ACM (2005) 1271–1272
8. Paolucci, M., Kawmura, T., Payne, T., Sycara, K.: Semantic matching of web services capabilities. In: First Int. Semantic Web Conf. (2002)
9. Dikenelli, O., Gümüs, Ö., Tiryaki, A., Kardas, G.: Engineering a multi agent platform with dynamic semantic service discovery and invocation capability. In Eymann, T., Klügl, F., Lamersdorf, W., Klusch, M., Huhns, M.N., eds.: MATES. Volume 3550 of Lecture Notes in Computer Science., Springer (2005) 141–152
10. Kardas, G., Gümüs, Ö., Dikenelli, O.: Applying semantic capability matching into directory service structures of multi agent systems. In: ISCIS. Volume 3733 of Lecture Notes in Computer Science., Springer (2005) 452–461
11. Zeng, L., Benatallah, B., Dumas, M., Kalagnanam, J., Sheng, Q.Z.: Quality driven web services composition. In: WWW '03: Proceedings of the 12th international conference on World Wide Web, New York, NY, USA, ACM Press (2003) 411–421
12. Cardoso, J., Sheth, A.P., Miller, J.A., Arnold, J., Kochut, K.: Quality of service for workflows and web service processes. J. Web Sem. **1** (2004) 281–308
13. Sacerdoti, E.: The nonlinear nature of plans. In: International Joint Conference on Artificial Intelligence. (1975)
14. Tate, A.: Generation project networks. In: International Joint Conference on Artificial Intelligence. (1977) 888 – 893
15. Erol, K., Hendler, J.A., Nau, D.S.: Complexity results for htn planning. Ann. Math. Artif. Intell. **18** (1996) 69–93
16. Sycara, K., Williamson, M., Decker, K.: Unified information and control flow in hierarchical task networks. In: Working Notes of the AAAI-96 workshop 'Theories of Action, Planning, and Control'. (1996)
17. Graham, J.R., Decker, K., Mersic, M.: Decaf - a flexible multi agent system architecture. Autonomous Agents and Multi-Agent Systems **7** (2003) 7–27

A Role-Based Support Mechanism for Service Description and Discovery*

Alberto Fernández, Matteo Vasirani, César Cáceres, and Sascha Ossowski

Artificial Intelligence Group, University Rey Juan Carlos
Calle Tulipán s/n, 28933 Móstoles (Madrid), Spain
{alberto.fernandez,matteo.vasirani,cesar.caceres,sascha.ossowski}@urjc.es

Abstract. The ever-growing number of services on the WWW provides enormous business opportunities. Services can be automatically discovered and invoked, or even be dynamically composed from more simples ones. In this paper we concentrate on the problem of service discovery. Most current approaches base their search on inputs and outputs of the service. Some of them also take into account preconditions and effects, and other parameters that describe the service. We present a new approach that complements existing ones by considering the types of interactions that services can be used in. We present our proposal for a concrete application based on a real-world scenario for emergency assistance in the healthcare domain.

1 Introduction

Service-Oriented Computing [9,10] is a software paradigm for distributed computing that is changing the way software applications are designed. Services are computational entities that can be described, published, discovered, orchestrated and invoked by other software entities. The Semantic Web Services Architectural framework [1] attempts to address the dynamic service discovery, service engagement, service process enactment and management, community support services, and quality of service.

Agent technology provides designers with an interaction-oriented way of designing open software systems [15]. There is a growing awareness that organisational models are fundamental to regulate open multiagent systems so as to instil desired properties [19,22,24]. Service-oriented and multiagent systems are quite related [11]. Often, web service technology is used to support the interactions in multiagent systems [5].

In this paper we concentrate on the problem of dynamic service discovery in multiagent systems. Most current service discovery techniques [14,17,20] aim at web services and base their search on inputs and outputs of the service. Some of them also take into account preconditions, effects and other parameters that

---

* This work has been partially supported in part by the European Commission under grant FP6-IST-511632 (CASCOM), and by the Spanish Ministry of Education and Science, projects TIC2003-08763-C02-02 and TIN2006-14630-C03-02.

J. Huang et al. (Eds.): SOCASE 2007, LNCS 4504, pp. 132–146, 2007.

describe the service. However, agent-based service discovery mechanisms can also make use of the information provided by the organisational model underlying the multiagent system. Following this idea, we present a new approach that extends existing mechanisms by considering the *types of interactions* that services can be used in. We have built a service matchmaker following this approach within the CASCOM project [2,8]. To describe our proposal we use examples from a concrete application based on a realworld scenario for emergency assistance in the healthcare domain.

This paper is organised as follows. In section 2 we present a role-based interaction modelling approach for service descriptions and illustrate it in the aforementioned use case scenario. In section 3 we outline what organisational information is relevant for service descriptions in multiagent systems, and show how it can be represented in OWL-S. The matchmaking algorithm is presented in section 4 and its implementation is evaluated in section 5. Finally, section 6 summarises our proposal and points to future lines of work.

2   Role-Based Interaction Approach to Service Description and Discovery

In order to improve both the efficiency and the usability of agent-based service-oriented architectures, we suggest exploiting common organisational concepts such as social roles and types of interactions to further characterise the context that certain semantic services can be used in. In this section we outline our role-based and interaction-centric modelling framework [18].

The CASCOM abstract architecture [3] conceives services to be delivered essentially by agents. In such an approach the agents usually act as mere wrappers for web services. The difference between a web service and a service provided by an agent boils down to a matter of interface: an agent can provide an implemented web service by a process of wrapping the service within an ACL interface in such a way that any agent can invoke its execution by sending the adequate (*request*) message.

However, agents are not only able to execute a service but can also engage in different types of interaction with that service. For example, consider a healthcare assistance scenario, similar to the one treated in the CASCOM project [4]: an agent providing a second opinion service should not only be able to provide a diagnostic; it may also be required to explain it, give more details, recommend a treatment, etc.

This means that the service provider is supposed to engage in several different interactions during the provision of a service. Thus, if a physician or a patient needs one or more second opinions, they should look for agents that include those additional interaction capabilities around the "basic" *second opinion* service. In a certain sense, this approach is similar to the abstraction that an object makes by providing a set of methods to manipulate the data it encapsulates. In this case, the agent provides a set of interaction capabilities based on the service.

Most current OWL-S matchmakers [17,14] only consider logical inference on service inputs and outputs in their matching algorithm. In the CASCOM project we use OWLS-MX [13], a hybrid semantic matchmaker that also uses information retrieval techniques for service matching. We have extended this matchmaker with role- and interaction-based matching techniques in order to adapt it to scenarios as the one outlined above.

In particular, the *efficiency* of the matchmaking process can be improved by previously filtering out those services that are incompatible in the terms of roles and interactions. The *effectiveness* of the matchmaking process can also be enhanced by including information regarding the roles and interactions. For instance, a diagnosis service may require symptoms and medical records as inputs and produce a report as output. However, the service functionality can be achieved either (i) by actually generating the report, (ii) by retrieving a previously done or (iii) by a brokering service to contact other (external) healthcare experts. As we will outline below, in all three cases the inputs and outputs are the same, but the role the service plays in the corresponding interactions is different.

2.1   Interaction Modelling

In order to develop role-based extensions to service discovery mechanisms we use a subset of the RICA organisational model described in [18] and [19]. Setting out from this basis, we first analyse different use case of the application domain scenario. For each use case, we identify the types of social interaction as well as the roles (usually two) that take part in that interaction. The next step is an abstraction process in which the social (domain) roles/interactions are generalised into communicative roles/interactions.

In the sequel, we will illustrate our approach based on the second opinion use case in the healthcare domain. In this scenario, the patient (or the physician of a local emergency centre) can ask an external health professional for a diagnosis on the basis of the symptoms and the medical records of the patient, like exams and past diseases.

The "conversation" between the patient and the health professional can be modelled by a sequence of (communicative) actions between the two agents involved, The patient *asks* the health professional for an opinion, providing the symptoms and the medical records. If there is insufficient information, the health professional *requests* additional information (possibly several times) and finally gives his *advisement*. If the provided diagnosis is not sufficiently clear, the patient can also solicit an *explanation*.

Starting from this conversation we can isolate 3 different interactions: (i) the second opinion exchange, which can comprise (ii) a detailed information exchange. When the second opinion exchange finishes, an explanation (iii) can occur.

The result of this analysis is a basic ontology of roles. Figure 1 shows an example, where the role *SecondOpinionRequestee* is generalized into a *Medical-Advisor* role, which in turn is generalized into an *Advisor* role. Similarly, the

SecondOpinion interaction can be generalized in a *MedicalAdvisement* interaction and then in an *Advisement* interaction, in which the Advisor informs the Advisee about his beliefs with the aim of persuading the Advisee of the goodness of these beliefs.

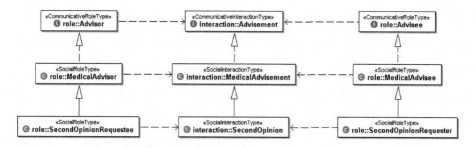

Fig. 1. Roles and interactions taxonomy

2.2   Role and Interaction Ontologies

From the use case scenarios, we have derived an ontology that contains a taxonomy of types of interactions, and a taxonomy of roles that take part in those interactions. Figure 2 shows a partial UML class diagram representation of the interaction type ontology. There are two kinds of interactions, social interaction types and communicative interaction types. *Social interaction* types are domain interactions (Emergency Medical Assistance, Medical Advisement, etc.). *Communicative interaction* types are generic reusable interaction patterns. They constitute abstract communication interactions that can be instantiated to different scenarios. For instance, a *medical advisement* is a specialisation, in the medical domain, of the generic *advisement* interaction type. In the UML diagram generic (communicative) interaction types are represented as interfaces, whereas domain (social) interactions are represented by classes.

The top concept of the ontology is the *Communication*, which represents the most generic type of interaction. Any type of interaction is a specialisation of this concept. The first level of specialisation is the *ClosedActionPerforming* interaction type, which represents any interaction that implies an action performing. *Advisement, Admission, Assistance* and *InformationExchange* are specialisations of *ClosedActionPerforming*. For example, an *information exchange* is a particular case of action performing, where the *informee* asks the *informer* to execute the action of informing about certain fact. Similar analysis can be applied to the domain interaction types that specialise communicative interaction types into social interaction types. For instance the *ArrivalNotification* is a particular kind of *notification* interaction about the arrival of a patient to a hospital. New (generic and domain) interaction types can be incrementally be added to this taxonomy making it more complete and reusable.

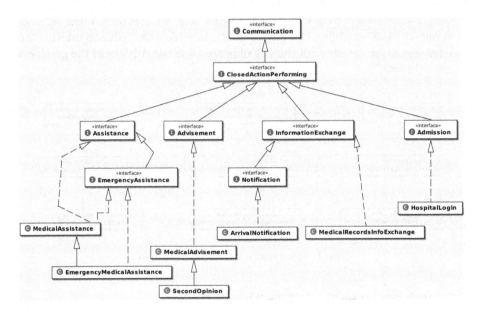

Fig. 2. Partial interaction type ontology

In line with the interaction taxonomy, there is a role taxonomy which includes a hierarchy of the roles that participate in the interactions. For each concept in the interaction ontology there will be at least two participant concepts in the role ontology. For example, the *informer* and *informee* are concepts of the role ontology that participate in the *information exchange* interaction.

These ontologies, and especially their generic (communicative) part, will be used in the service description and matchmaking extensions.

3   Role-Based Service Description

In this section we outline our proposal to integrate role-based information into service descriptions. We first describe a representation for role-based service advertisements and service requests, and then explain how this information can be included in the OWL-S service profiles, the service description language we use.

3.1   Service Advertisements

Role-based service descriptions comprise two kinds of information related to the interactions in which the service provider agent can engage:

1. the *main role* played in the interaction, e.g. the *advisor* role in the second opinion service;
2. a set of roles that may be *necessary* to be played by the requester for the correct accomplishment of the service. For instance, in an advisement interaction of a second opinion service, the provider may need to initiate an

information exchange interaction in which it plays the *informee* role, and the requester plays the *informer* role. Necessary roles are given by a formula in disjunctive normal form, i.e. a disjunction of conjunctions of roles.

These two fields are repeated for each main role the service can play. We may graphically represent a service advertisement by a table with two rows, in which each column contains the main role (first row) and the necessary roles (second row).

Table 1. Abstract example of a role-based service advertisement

Main Role	Necessary Roles
R_1	$(R_4 \wedge R_5) \vee R_6$
R_2	$(R_6 \wedge R_7 \wedge R_8) \vee (R_4 \wedge R_7)$
R_3	

In this abstract example, the service provider agent can engage in three different types of interactions when providing the service. It can play role R_1, which requires the requester to be able to play both R_4 and R_5, or R_6. If this condition fails, then the provider will not be able to carry out the service properly. However, when the provider plays the role R_2, it requires the requester to play either roles R_6, R_7 and R_8, or otherwise roles R_4 and R_7. In the case of R_3, no role playing capabilities are mandatory for the requester.

3.2   Service Requests

In the case of a service requests, we consider that a query comprises two elements:

1. *Main roles* searched. Although one role will be enough in most cases, we allow for more complex search patterns in which the provider is able to play more than one role. As in the case of service advertisements, we require an expression in disjunctive normal form here.
2. A set of roles that define the *capabilities* of the requester. These are roles the requester is able to play. This information is important if the provider requires interaction capabilities from the requesters. For example, the requester of a second opinion can inform that it is able to provide information (*informer*) if needed.

Table 2 shows an abstract example of a query in which the requester searches a service provider able to play either R_2 or both R_4 and R_6 to provide the service. In addition, it states that it is able to play the roles R_4, R_7 and R_1 if required. The service described by Table 1 matches this query because the provider is able to play one of the roles (R_2) required by the requester and the requester can play some roles (R_4 and R_7) that make true the necessary role conditions.

Table 2. Abstract example of a role-based service request

Main Roles	$R_2 \vee (R_6 \wedge R_4)$
Capabilities	R_4, R_7, R_1

More formally, we describe a registered service advertisement S and a service request R as:

$S = SET\ OF\ InteractiveRoleDescription$
$InteractiveRoleDescription = \langle MainRole, RoleExpression \rangle$
$RoleExpression = SET\ OF\ ConjunctiveRoleList$
$ConjunctiveRoleList = SET\ OF\ Roles$

$R = \langle MainRoleR, Capabilities \rangle$
$MainRoleR = RoleExpression$
$Capabilities = SET\ OF\ Roles$

Tables 3 and 4 show a role-based service advertisement and request, respectively, for the second opinion example. The request specifies that the requester is able to play the informer and explainer roles if necessary.

Table 3. Second Opinion role-based service advertisement

Main Role	Necessary Roles
Advisor	Informer
Explainer	
Informer	

Table 4. Second Opinion role-based service request

Main Roles	Advisor $\wedge$ Explainer
Capabilities	Informer, Explainer

Notice that our approach is compatible with services that do not make use of the role-based extensions in their description. In case a service description does not include the role-based approach, we assume it has a main role *Communicator* (the top and most general concept of the ontology) and no necessary roles are required from the requester. If the request does not include a role description, we assume the requester is not interested in the role-based approach and the matchmaker will omit that phase in the service matching process.

3.3   Extending OWL-S with Roles

We use the service description language OWL-S [16] in our system. An OWL-S service description consists of three parts: the *service profile* used for advertising and discovering services; the *process model*, which details how the service operates; and the *service grounding*, which provides details on how to interoperate with the service.

The service profile is designed to describe what the service does, so this is the adequate location for role descriptions. A service profile basically includes information about inputs, outputs, preconditions and effects, as well as non-functional information like service category, service classification, service product, etc. However, organisational information such as roles or interactions does not fit in any of those predefined fields, so we opt for including the role description as an additional parameter, called *ServiceRoles* in the case of service descriptions and *QueryRoles* for service requests.

4   Role-Based Service Matching Algorithm

We have developed a role-based matching algorithm that takes as inputs a service request (R) and a service advertisement (S), and returns the degree of match (*dom*) between them. Essentially, it searches the role in the advertisement S that best matches the one in the query (R). The match between a role in the query and one role in the advertisement is made by the function *MatchAtomicRequest*, which also receives the set of capabilities of the requester, in case it needs to check if the necessary roles can be provided by the requester.

Figure 3 shows a pseudocode for the matching algorithm (*Match* function). As described before, the request may not only include a role but also an expression (a disjunction of conjunction of roles). The loops in lines 4 and 6 decompose that expression, using the minimum as combination function for the values in a conjunction and the maximum for disjunctions.

MatchAtomicRequest returns the degree of match between a role in the request and a service advertisement, given the set of capabilities of the requester. It compares the requested role with every other given role and returns the maximum degree of match. For each role in the advertisement, the match between the main roles is made, as well as the match between the necessary roles and the capabilities of the requester. The minimum of both values is considered the degree of match. Again, the necessary roles are given by a logical expression, which must be evaluated by decomposing it within the *MatchRoleExpr* function.

The semantic match of two roles R_A (*advertisement*) and R_Q (*query*) is made based on the ontology of roles. It is a function that depends on two factors:

1. Level of match. This is the (subsumption) relation between the two concepts (R_A, R_Q) in the ontology. We differentiate among the four degrees of match proposed by Paolucci et al. [17]:
   (a) *exact* : if $R_A = R_Q$
   (b) *plug-in* : if R_A subsumes R_Q
   (c) *subsumes* : if R_Q subsumes R_A
   (d) *fail* : otherwise
2. The distance (number of arcs) between R_A and R_Q in the taxonomy.

We consider, for the moment, that all roles have the same importance and that the generality (depth in the taxonomy) of the roles is not relevant. We combine both criteria into a final degree of match which is a real number in the range [0, 1], so service providers can be selected by simply comparing these numbers. In this combination, the level of match always has higher priority: the value representing the degree of match is equal to 1 in case of an *exact* match, it varies between 1 and 0.5 in case of a *plug-in* match, rests between 0.5 and 0 in case of a *subsumes* match, and it is equal to 0 in case of a *fail*.

There are infinite functions that fulfil that precondition. One equation that implements this behaviour is that in equation 2, where x is the distance between R_A and R_Q ($depth(R_A) - depth(R_Q)$) in the role ontology (if there is a subsumption relation between them). This kind of function guarantees that the value of a *plug-in* match is always greater than the value of a *subsumes* match, and it only considers the distance between the two concepts, rather than the total depth of the ontology tree[1], which may change depending on the domain. Furthermore, the smaller the distance between concepts (either in the case of *plug-in* or *subsumes* match), the more influence will have a change of distance in the degree of match.

Consider, for instance, the example case described in section 3. The degree of match between the advertised role *Advisor* and the required role *SecondOpinionRequestee* (that are related by a subsumption relation like the interactions *Advisement* and *SecondOpinion*) is:

$$x = depth(Advisor) - depth(SecondOpinionRequestee) = 4 - 2 = 2$$

$$dom(x) = \tfrac{1}{2} + \tfrac{1}{2 \cdot e^2} = 0.5677$$

(1)

The necessary role *Informer* is included in the capabilities set of the requester, so an exact match (1) is obtained. The final degree of match is the minimum of both values, i.e., 0.5677, which corresponds to a *plug-in* match.

$$dom(R_A, R_Q) = \begin{cases} 1 & if \ R_A = R_Q \\[2mm] \frac{1}{2} + \frac{1}{2 \cdot e^{\|R_A, R_Q\|}} & if \ R_A \ is \ subclass \ of \ R_Q \\[2mm] \frac{1}{2} \cdot e^{\|R_A, R_Q\|} & if \ R_Q \ is \ subclass \ of \ R_A \\[2mm] 0 & otherwise \end{cases}$$

(2)

---

[1] Note that, for instance, if a lineal function is used, the maximum possible distance between two concepts must be known a priori to establish the equation (e.g. $dom(x) = 1 - x/6$).

```
01 Match(R: service request, S: service advertisement)
02 {
03   dom = 0
04   FOR ALL CRLi IN R.MainRoles {
05     dom' = inf
06     FOR ALL rj IN CRLi {
07       dom' = min(dom',MatchAtomicRequest(rj,R.Capabilities,S))
08     }
09     dom = max(dom, dom')
10   }
11   return dom
12 }
13 MatchAtomicRequest(role: Role, Capabilities: SET OF ROLES,
14                   S: service advertisement)
15 {
16   dom = 0
17   FOR ALL IRDi IN S.InteractiveRoleDescription {
18     dom1 = MatchRole(role,IRDi.MainRole)
19     dom2 = MatchRoleExpr(IRDi.RoleExpression, Capabilities)
20     dom = max(dom, min(dom1,dom2))
21   }
22   return dom
23 }
24 MatchRoleExpr(RExpression: SET OF ConjunctiveRoleList,
25               Capabilities: SET OF Roles)
26 {
27   dom = 0
28   FOR ALL CRLi IN RExpression {
29     dom' = inf
30       FOR ALL rj IN CRLi {
31         dom' = min(dom',MatchRoleInList(rj, Capabilities))
32       }
33     dom = max(dom, dom')
34   }
35   return dom
36 }
37 MatchRoleInList(role: Role, RL: SET OF Roles)
38 {
39   dom = 0
40   FOR ALL ri IN RL {
41     dom = max(dom, MatchRole(role,ri))
42   }
43   return dom
44 }
```

Fig. 3. Matching algorithm

5   Evaluation

We have constructed a matchmaker (named ROWLS) that implements the algorithm described in this paper. Our implementation relies on the Mindswap Java Library [2] for parsing OWL-S service descriptions, and on Jena [3] for managing OWL ontologies.

We have realized several experiments to evaluate the efficiency (response time) and the effectiveness of our approach. For our testing, we used a subset of the OWLS-TC v2 [4]. We annotated selected service descriptions with organisational information (roles) and added new services and queries, leading to a set of 378 OWL-S that were used in our experiments.

As described before, the approach presented in this paper is intended to be complementary to other general-purpose matchmakers. In our experiments, we combined ROWLS with OWLS-MX [13], one of the leading hybrid matchmakers available to-date. The results reported in the next subsections mainly compare relevant features of a combination of ROWLS and OWL-MX compared to a standalone use of the latter.

5.1   Efficiency Evaluation

Figure 4 depicts the results of the scalability tests that we have performed to evaluate ROWLS[5]. It shows that the matching time increases linearly with the number of services, giving an average of 0.03 ms in matching a query with one service. This is two orders of magnitude less than OWLS-MX matching time. We explain this by the fact that ROWLS concentrates on a smaller number of characteristics of a service description. In addition, the ontology used by ROWLS is smaller than the domain ontologies that OWLS-MX has to deal with.

Load time was not considered in the scalability test because it depends on several external factors like disk access time, network connections, web servers, etc. Note that parsing a service description usually implies accessing (via Internet) and parsing several domain ontologies. Our tests revealed an average load time between 100 and 200 ms per service, which is much higher than match time per service. This implies two important architectural issues: (i) the combination of both matchmakers should be tight, sharing the internal representation of the services (i.e. only loading every service once), and (ii) it is recommended to integrate the matchmakers within the directory such that service descriptions are parsed only once for different queries by implementing some sort of caching mechanism.

5.2   Effectiveness Evaluation

Another relevant question is as to how far the use of ROWLS can improve the results of a general-purpose matchmaker in terms of effectiveness. In particular,

---

[2] www.mindswap.org/ mhgrove/kowari

[3] http://jena.sourceforge.net

[4] http://projects.semwebcentral.org/projects/owls-tc/

[5] We used an Intel Pentium 4 3.00GHz computer, with 1GB Ram.

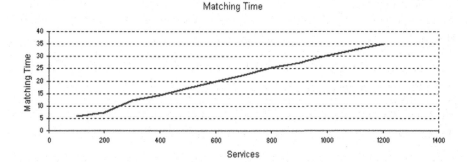

Fig. 4. Matching time (ms)

we were interested in measuring the performance of an architecture in which ROWLS is used as a filter which eliminates a number of services, so that they need not be fed as input to OWLS-MX. For this purpose, our tests were carried out with different ROWLS filter configurations that pass on only a certain percentage of the best-ranked services (according to role-based matching) to OWLS-MX (30%, 60%, 90% and 100%, respectively).

Our first experiments shed light on the different relations between *precision* and *recall* [21]. Figure 5 shows the (macro-averaged) precision-recall-curves for OWLS-MX alone and for the four combinations of ROWLS and OWLS-MX. As shown in the figure, based on our test collection, using ROWLS as a filter for the general-purpose OWLS-MX matchmaker yields to better precision compared to a standalone version of OWLS-MX for all level of recall.

In open multiagent systems with thousand of services, which is the target domain of our research, precision becomes more relevant than recall: we are interested in finding one (or a small set of) service providers, but it is not necessary to retrieve all of them. To account for this fact, we have performed experiments

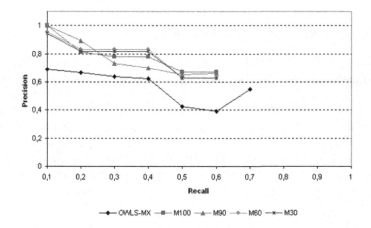

Fig. 5. Precision-recall curves

to come up with quantitative data regarding two alternative measures of effectiveness proposed within TREC[6] in such situations, namely *average precision* and *R-precision*.

The *average precision* is shown in Figure 6 (left). It can be seen that the combination of OWLS-MX and ROWLS outperforms a standalone OWLS-MX based on our test collection: the less services pass the filter (ROWLS) the higher is the precision. This is because ROWLS' role-based matching is orthogonal to the general-purpose matching strategy of OWLS-MX, thus filtering out irrelevant services that OWLS-MX would have erroneously classified as relevant.

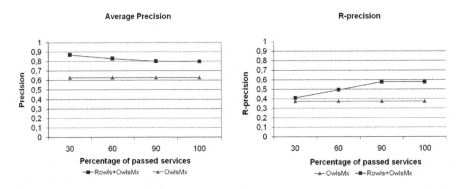

Fig. 6. Average precision and R-precision

Figure 6 (right) shows the *R-precision* values based on our test collection[7]. Again, ROWLS enhances the behaviour of OWLS-MX. Note that the R-precision increases with the number of services that pass the filter. This may happen because only the best ranked services pass the ROWLS filter, and the more services are fed into OWLS-MX, the higher the probability that irrelevant services are among them.

6   Conclusions and Future Lines

In this paper, we have described a novel approach for service description and discovery in multiagent systems based on organisational models. We have identified roles and interaction ontologies as key information to this respect, have applied them to service advertisements and requests, and have formalised both on the basis of OWL-S. An extension to common service matching techniques that exploits this additional organisational information has been put forward, which has given rise to the implementation of the ROWLS role-based matchmaker. We have shown that using ROWLS as a filter for a general-purpose matchmaker such as OWLS-MX may enhance the overall performance of the matchmaking process.

---

[6] http://trec.nist.gov/
[7] Precision after retrieving R services, being R the number of relevant services of the test collection.

Organisational concepts, such as roles and interactions [12], are abstraction mechanisms that have been used for many years in object oriented models, and are now usually present in agent-oriented design methodologies [22,6,7]. However, current service discovery approaches [14,17,20] do not make use of these abstractions to describe and search service providers. To the best of our knowledge there are no service discovery approaches that exploit role and interaction type information in multiagent systems.

We are considering some variations in several aspects of the algorithm presented in order to improve its precision. In particular, we will study how the matchmaker behaves if other aggregation functions are used instead of minimum and maximum to combine the degrees of match of individual roles, considering methods from the field of fuzzy logic such as more sophisticated t-norms and t-conorms.

References

1. Burstein, M., Bussler, C., Zaremba, M., Finin, T., Huhns, M. N., Paolucci, M, Sheth, A., Williams, S.: A Semantic Web Services Architecture. IEEE Internet Computing 9(5) (2005) 72-81

2. Cáceres, C., Fernández, A., Ossowski, S.: CASCOM - Context-aware Health-Care Service Coordination in Mobile Computing Environments. ERCIM News 60 (2005) 77-78

3. Cáceres, C., Fernández, A., Ossowski, S., Vasirani, M.: The CASCOM Abstract Architecture for Semantic Service Discovery and Coordination in IP2P environments. In 8th International Conference on Enterprise Information Systems, Paphos - Cyprus (2006).

4. CASCOM, (2005). http://www.ist-cascom.org

5. Cavedon, L., Maamar, Z., Martin, D., Benatallah, B.: Extending Web Services Technologies: The Use of Multi-Agent Approaches. Springer (2004)

6. DeLoach, S.A., Wood, M. F., Sparkman, C. H.: Multiagent systems engineering. Int. Journal of Software Engineering and Knowledge Engineering, 11(3) (2001) 231-258

7. Giorgini, P., Kolp, M., Mylopoulos, J.: Multi-Agent Architectures as Organizational Structures. Journal of Autonomous Agents and Multi-Agent Systems. Kluwer Academic Publishers (2003)

8. Helin, H., Klusch, M., Lopes, A., Fernández, A., Schumacher, M., Schuldt, H., Bergenti, F., Kinnunen, A.: Context-aware Business Application Service Coordination in Mobile Computing Environments. In AAMAS05 workshop on Ambient Intelligence - Agents for Ubiquitous Computing, Utrecht (2005)

9. Huhns, M. N., Singh, M. P.: Service-Oriented Computing. John Wiley & Sons (2005)

10. Huhns, M. N., Singh, M. P.: Service-Oriented Computing: Key Concepts and Principles. IEEE Internet Computing, 9 (1) (2005)

11. Huhns, M. N., et al: Research Directions for Service-Oriented Multiagent Systems. IEEE Internet Computing, 9 (6) (2005)

12. Karageorgos, A: Using Role Modelling and Synthesis to Reduce Complexity in Agent-Based System Design. In Dept. of Computation, doctorate thesis, University of Manchester, Institute of Science and Technology, Manchester, (2003)

13. Klusch, M., Fries, B., Khalid, M., Sycara, K.: OWLS-MX: Hybrid Semantic Web Service Retrieval. Proceedings 1st International AAAI Fall Symposium on Agents and the Semantic Web, Arlington VA, USA (2005)
14. Li, L., Horrock, I.: A software framework for matchmaking based on semantic web technology. In Proc. 12th Int World Wide Web Conference Workshop on E-Services and the Semantic Web (ESSW) (2003).
15. Luck, M., McBurney, P., Shehory, O., S. Willmott: Agent Technology: Computing as Interaction (A Roadmap for Agent Based Computing). AgentLink, (2005).
16. OWL-S Home Page. http://www.daml.org/services/owl-s/
17. Paolucci, M., Kawamura, T., Payne, T., Sycara, K.: Semantic matching of web services capabilities. In Proceedings of the First International Semantic Web Conference on The Semantic Web, Springer-Verlag (2002) 333-347
18. Serrano, J.M., Ossowski, S., Fernández, A.: The Pragmatics of Software Agents - Analysis and Design of Agent Communication Languages. Intelligent Information Agents - The European AgentLink (Klusch et al. ed.), Springer (2002) 234-274
19. Serrano, J.M., Ossowski, S.: A computational framework for the specification and enactment of interaction protocols in multiagent organizations. To appear in: Journal of Web Intelligence and Agent Systems, Idea Press (2006)
20. Sycara, K., Klusch, M., Widoff, S., and Lu, J.: Larks: Dynamic matchmaking among heterogeneous software agents in cyberspace. Journal of Autonomous Agents and Multi-Agent Systems, 5(2). Kluwer Academic Press. (2002)
21. Van Rijsbergen, C. J.. Information Retrieval, 2nd edition. Dept. of Computer Science, University of Glasgow, 1979.
22. Wooldridge, M., Jennings, N. R., and Kinny, D.: The Gaia Methodology for Agent-Oriented Analysis and Design. Journal of Autonomous Agents and Multi-Agent Systems 3(3) (2000) 285-312.
23. WSMO working group http://www.wsmo.org/.
24. Zambonelli, F., Jennings, N. R., Wooldridge, M.: Organizational Abstractions for the Analysis and Design of Multi-agent Systems. Agent-Oriented Software Engineering: First International Workshop, AOSE 2000, Limerick, Ireland, June 10 (2000) 235-251

WS2JADE: Integrating Web Service with Jade Agents

Xuan Thang Nguyen and Ryszard Kowalczyk

Swinburne University of Technology, Faculty of Information and Communication
Technologies, Melbourne VIC 3122, Australia
{xnguyen,rkowalczyk}@ict.swin.edu.au

Abstract. Web services have gained popularity today for enabling universal interoperability among applications. In many scenarios, allowing software agents to access and control Web services is important and hence the integration between these two platforms. In this paper, we focus on technical aspects of an integration framework of Web services and Jade agent platform. The mismatch of description and communication between FIPA compliant agent platforms versus Web services are two key challenges that must be addressed. Our implementation, WS2JADE, is described and compared with WSDL2JADE - a previous implementation on the same topic, and WSIGS - a recent proposal of Web service and Agent integration architecture. In contrast to other solutions, WS2JADE provides facilities to deploy and control Web services as agent services at run time for deployment flexibility and active service discovery.

1   Introduction

With the emergence of the Web services standards, universal interoperability between applications is becoming reality. Following a loosely coupled integration model and common service access standards, Web services enable flexibility in the integration of heterogeneous systems. However, how to achieve the automation of Web services discovery, composition and invocation, or how to perform Web service execution monitoring and management are still open issues. Agent software is a promising technology to solve such problems. Unfortunately, Web services and agents were originally developed separately with different standards and specifications. Therefore, integrating between these two platforms becomes important in this context.

There has been many research on the topic of Web services and agent integration, to provide access to Web services from agent platforms and vice versa. This issue has also been addressed in the AgentCities project [3]. Main identified obstacles are the description mismatch and communication mismatch between Web services and Agents. A proxy approach has been recommended as the most suitable solution for the current stage of Web services and FIPA compliant agent platforms.

In this paper, we propose an enhanced solution for Web services and FIPA compliant Multi Agent System integration that offers many advantages over existing solutions. Specifically, we discuss different ways how Web services can be

J. Huang et al. (Eds.): SOCASE 2007, LNCS 4504, pp. 147–159, 2007.

visible to Agents and how they can be accessed and used by Agents. We first present some background on the issue and then propose an approach for the integration followed by a proof-of-concept implementation for a specific case of Web services and JADE agent platform. We contrast our tool with WSDL2JADE and WSIGS to show that WS2JADE, compared to WSDL2JADE, has more advanced features including dynamic Web Service deployment at runtime, improvements of more capable proxy agents and WSDL parsing functionality.

In the next section the paper discusses the related work. The WS2JADE approach is proposed in Section 3, followed by an example of its functionality in Section 4. Finally, Section 5 concludes the paper and outlines future work.

2   Related Work

In the area of theoretically-related work, a symmetric integration of Web services and FIPA-compliant agent platforms has been proposed in [3] as a high-level architectural recommendation from the AgentCities. The reason of this symmetry is that Web services were developed without the concept of Agents (i.e. FIPA agents) and can exist without Agents. The symmetric architecture takes into consideration that many Web Service clients, though may have autonomous characteristics of agents, are not conformed to the FIPA specifications. A proxy-based approach allows the two platforms to be evolved in parallel without imposing any restrictions on each other. This approach accepts the equity between the roles of agents and Web services, which is different to the traditional view that Agents are considered one level up from Web services, and agents take solely the roles of Web services providers and consumers. As can be seen from Figure 1 taken from [3], a FIPA agent service environment exists in parallel with a Web service environment. The "FIPA Agent Service to Web service Gateway" on the border between the two environments allows FIPA agents to access Web services

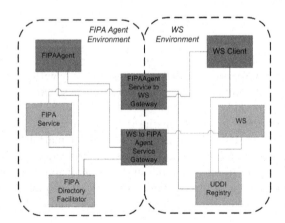

Fig. 1. Agents and Web services can communicate through language translation gateways [7]

by translating ACL messages to Web service invocations. In the reverse order, the "Web Service to FIPA Agent Gateway" exposes and registers agent services in UDDI Registry Server so that any Web service client can use them. Following AgentCities recommendations, two separate pioneer implementations have been proposed to solve two ends of the problems for FIPA compliant Jade Agent system: Exposure of Web services to Jade agents by Sztaki [12] and exposure of Jade agent services to Web services by Whitestein Technology [1]. Whitestein Technology has released their tool WSAI (Web service Agent Integration) as an open source code in its first version. Another tool from Whitestein, WSIGS (Web Service Integration Gateway Service), is under development and its architecture has been published in [11]. WSDL2JADE has been released as an online program that converts WSDL file to Jade classes. It takes an input of a Web service address and generates outputs of Jade agent code and agent ontology for the Web service. There is no run-time deployment capability. Based on some test cases we have carried out with Sztaki's WSDL2JADE, there are also some problems we notice with that tool. A sample of input WSDL test file can be found at WSDL2JADE homepage [2]. For example some XML enumeration values and type information are missing in the generated ontology files with XML Enumeration data type. It prevents client agents to use operations related to this data type. WSAI and WSIGS have been proposed by Whitestein Technology [11]. WSAI [1] allows Web service clients to use JADE agents' services. In order to do this, WSDL files are generated for these agent services. Technically, at this stage WSDL files are created manually from these agents' behaviors. It also requires "interface agents" to communicate with a target agent. These "interface agents" are created and destroyed per Web service client invocation of the agent service. However it appears that the applicability of WSAI in practical situations is limited because of two major obstacles. Firstly, Jade agent service specification is specified loosely and not based on message levels. This makes it difficult to automatically generate WSDL interfaces. Secondly, the default single-threaded mode of Jade agent, and the asynchronous and stateful nature of agent communication do not fit well in the current implementation stage of Web service communication which mostly focuses on synchrony. There have been discussions of asynchrony versus synchrony in [3]. However, how to translate stateful communications of Agents, in which conversations and history of past interactions with other Agents are remembered, to stateless communications of Web service is not discussed. We believe that the WSRF (WS-Resource Framework) specification [9] where a stateless service can have stateful resources could be applied here for a solution.

WSIGS is under development at the writing time of this paper. WSIGS proposes an architecture for bi-directional integration with no special Agents [11]. WSIG is a set of codecs that do the translation between agents' ACL (Agent Communication Language) and Web Services' calls. To be visible in both environments, WSIG is registered as a special agent service in FIPA DF (Directory Facilitator) and a special Web Service endpoint in UDDI directories. When an agent wants to invoke a service (Web service) registered in WSIG registry, the request is passed on to WebServiceInvocation, a component of WSIG, to perform the actual Web service invocation. The requirement for services in one environment to be registered by their owners in a public directory before they can be seen in other environments reflects the assumption that Web services need to be registered before they can be discovered. This is true for a model like UDDI but in more recent P2P models the assumption is no longer hold.

3   WS2JADE Approach

This section describes a proposed approach for the integration of Web services and Jade agents with WS2JADE. From an architectural perspective, WS2JADE, in accordance with [3], forms a Web service to FIPA Agent Service gateway. There are two distinct layers in WS2JADE: interconnecting layer and management layer. The interconnecting layer contains interconnecting entities that glue Web Service and agents together. The management layer, being static, creates

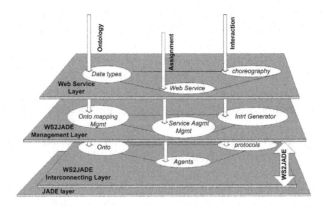

Fig. 2. Web services are connected to JADE agents via a management layer and a communication interconnecting layer

and manages those dynamic interconnecting entities. In WS2JADE, the interconnecting entities consist of special agents, ontology and protocol specifications. We call these special agents WSAG (Web Service Agents). WSAG are the agents capable of communicating with and offering Web Services as their own services. The combination of a static and a dynamic layer is a distinct feature of WS2JADE as compared to different tools mentioned in the previous part. The WS2JADE static management layer is capable of active service discovery and automatically generating and deploying WSAG at runtime. It is an advancement over WSDL2JADE since it can automate the agent deployment process. Instead of being passive like WSIG, the WS2JADE is designed to actively discover Web services and registers them on DF if required. Also, each WSAG is multi-threaded and has its own Web service invocation module.

Figure 2 shows that WS2JADE management can be looked at from a different perspective as a layer which is capable of projecting Web Service or any

Service Oriented environment layer into JADE agent layer. The output of this projection is the interconnecting entities. As depicted in this figure, three mappings are carried out by WS2JADE during the projection: ontology mapping, interaction mapping, and assignment mapping. These mappings are handled by three main components that construct the WS2JADE management layer: ontology generator and management, interaction generator and management, and service assignment management. These components form the WS2JADE management part as shown in figure 3. Figure 3 presents different components within

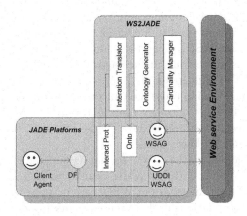

Fig. 3. Different system components in WS2JADE

WS2JADE system and how they are linked to JADE. The vertical rectangular box is WS2JADE, the horizontal one is JADE. Note that the overlap between WS2JADE and JADE is the components in WS2JADE interconnecting layer: generated interaction protocols, ontologies, and WSAGs. Figure 3 also illustrates a scenario for WS2JADE operation, in which a client agent searches for some service on DF. The DF can trigger WS2JADE to look up for available services in the Web service environment. If some Web services are found, their corresponding ontology and interaction models are generated. Also, a WSAG capable of accessing the Web service is generated. This WSAG registers the Web service as its service on DF and communication between the client agent and this WSAG can start if the client agent wants the service. The following will discuss each WS2JADE component in turn.

3.1   Ontology Generation and Management

The ontology generator is responsible for ontology generation and ontology management. It translates data and its structure from Web service WSDL interfaces into meaningful information for Agents. A detailed explanation of a WSDL document can be found in WSDL specification [8]. WSDL describes abstract concepts and concrete entities. Abstract concepts are port type, operation, message and data type. Concrete entities are data encoding style, transport protocol and network address. In WS2JADE, the abstract concepts are relevant for the ontology mapping management as Agents need to know how to invoke operations of a Web service. The concrete entities are handled by the interaction translator and management component. WS2JADE ontology translator and management component converts Web services' data types and operation inputs and outputs into agent ontologies. The corresponding WSDL port type is tagged in the structure of the ontologies. According to JADE documents [1], JADE ontologies can be represented as Java classes, which are convenient for Jade agent's manipulation and processing. Alternatively, it can be in other formats such as RDFS and OWL for interoperability with other FIPA compliant agent platforms. Our WS2JADE toolkit supports Jade native ontology and OWL. To generate ontologies in Java, a WSDL data type is converted to a concept in agent ontologies. Two concepts are generated for each WSDL operation. One is for the operation input message and the other is for the output message. WSDL data types can be built-in XML types. The list of built-in simple XML data types are defined in the XML schema specification. We map these built-in simple data types to Java primitives that are supported by Jade ontology representation. For XML data types that are not built-in, special customized Java classes are used, for examples, Beans, Enumeration Holders and Facet classes.

Generating ontologies in the OWL format is simpler than in Java classes because OWL and WSDL both use XML. Similar to Jade ontology generation approach, data types and messages in WSDL are mapped to concepts in OWL. There is a one-to-one relationship between concepts in ontologies generated in Java and OWL . OWL is still very new and subjected to changes; however we share the belief that it will continue to play an important role in Semantic Web with an increasing support from agent communities.

In addition to ontology generation, ontology management is important in WS2JADE. WS2JADE organises generated ontologies in an efficient way. For data types that can be shared among different Web services, the corresponding generated ontology concepts are shared and form a common ontology base. It means that every time a new Web service is presented as an agent service, part of the existing ontology base and domain knowledge can be reused for this new service. Also, this allows the ontologies to be structured in a manageable way.

3.2   Interaction Mapping

The interaction mapping management component handles the conversion from Web Service communication into agent communication. Specifically, it converts Web service transport messages into ACL envelopes and Web service interaction patterns into agent protocols. These are correspondent to two sub-functionalities: language translation and interaction conversion.

Language Translation: To translate Web Service transport messages (commonly SOAP) into agent ACL messages, the SOAP envelope is first projected into Java languages and then into ACL. We do not translate SOAP into ACL directly because of two reasons. Firstly, we want to reuse our generated ontologies and existing Java implementations of SOAP. Secondly, we make an assumption

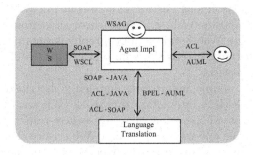

Fig. 4. Interaction Translator and Management

of the agent's intelligent capabilities to understand and process the messages according to its own logic in addition to language translation before forwarding the messages. This is best done by translating SOAP and ACL into Java - the native language for JADE. Figure 4 indicates that when a WSAG receives a SOAP message, it uses the Language Translation component to convert this message directly to ACL and send to the client agent. It can also perform some reasoning and modification on the message by converting the message to Java classes before any translation into ACL. In the Language Translation part of the interaction management component, Axis' JAX-RPC (Java API for XML based Remote Procedure Call) implementation and JADE support for content languages and ontologies are used for translations between XML and Java, and between ACL and Java. Axis is one of the most popular open source implementations of SOAP today. On one hand, JAX-RPC, led by Sun, is a specification of Web Service Invocation framework in Java. In JAX-RPC specification, at the client side, Java to XML translation in remote method call is done through a mapping from Java client stubs to the SOAP message representation. On the other hand, in JADE, information represented in Jade ontology-supported classes (Java objects) can be converted to different ACL content languages, including SL and LEAP. As can be seen from Figure 4, language translation is leveraged by the reuse of existing technologies instead of reinventing the wheel. SOAP-ACL translation is done by piping SOAP-JAVA and ACL-JAVA translation together. The main task of the language translation component is to map Axis stubs to Jade ontologies. However, we found out that due to the restrictions of Jade ontology and JAX-RPC classes it is not easy to convert data between them. In particular, an automation of the conversion process for any data types is difficult. We use special classes which represent the ontology facets to preserve precisions in the conversion process. There has been a similar discussion in [12] for Sztaki WSDL2JADE. Complex data mapping in WS2JADE (for example mapping of Axis Holder and Enumeration types to Jade ontology concepts and classes) is done recursively through simple data type.

Interaction Pattern Translation: In the interaction pattern translation component, we focus on choreography. By "choreography" we mean the required patterns of interactions among parties. It is in contrast to "orchestration" that describes how a composite Web service is constructed from other atomic services. For a composite Web Service, choreography is obtained by looking from an outsider's perspective. It tells the Web service clients different steps of how to use a composite service.

We have mapped simple interactions implicitly described in WSDL documents into standard FIPA interaction protocols. Web service (WSDL version 1.2) provides four types of operation: one-way, request-response, solicit-response, and notification. In the one-way operation, a Web service client sends a request without receiving any response from the Web service. In the request-response, the client sends a request and receives a response synchronously. In the solicit-response, the Web service sends a solicit request to the client and receives a response. In the last type, notification, the Web service notifies the client without receiving any response. These four types of Web service operations lead to three common interaction patterns in practice: request-response, solicit-response, and subscribe-notification. The request-response and solicit-response interaction patterns correspond to those of Web service operation types. The subscribe-notification interaction describes the conversation style in which a client registers to the Web service in order to receive notifications when some event occurs. Table 1 summaries the mapping between these interactions styles into agent protocols. More information on FIPA Request Interaction protocol, Request Interaction protocol, and Subscribe Interaction protocol can be found in [5].

3.3 Service Assignment Management and Service Discovery

The Service Assignment Management component is responsible for cardinal mapping and service deployment management. The cardinal mapping management manages M:N relationship between Web services and WSAG. In WS2JADE, a number of Web services can be offered as services of different WSAGs. In other words, a WSAG can offer more than one Web Service and a Web Service can be offered by more than one WSAG. This cardinal relationship is managed through a registry that keeps records as triples of a Web service, a WSAG that offers the Web service, and a new name of the Web Service in the agent platform. The service assignment management also provides a tool for deploying and destroying WSAGs, and assigns new Web services to a WSAG. It informs the WSAG which ontologies should be used for a particular newly assigned Web Service. If an assigned Web service is reported to be no longer available, the service deployment

Table 1. Interaction pattern Mapping

WS Interaction patterns	Agent protocols
Request-response	FIPA Request Interaction Protocol
Solicit-response	FIPA Request Interaction Protocol
Subscribe-Notification	FIPA Subscribe Interaction Protocol

management removes the service from the list of the offered services of WSAGs and from the DF.

The Service Discovery component is designed to discover Web services. It is essentially a piece of software that can use Web service discovery protocols and translate the received information into agent service description for DF. As mentioned above, we prefer an active discovery model rather than waiting for services to be registered. At the time of this writing, Web service discovery protocol is complex and subject to change with the latest revised version of WS-Discovery specification which uses multicast protocols. Traditional Web service discovery mechanism of UDDI shares a common model with agent DF in a sense of accessing the directory. However, UDDI has evolved away from the concept of a "Universal Business Directory" that represented a master directory of publicly available services as DF still is. Most P2P based and multicast discovery protocols prove that requesting service providers to register the services is not always the case.

3.4   Remarks

As discussed in [3], the main difficulties in integration of Web services and FIPA compliant agent platforms are mismatches in communication and descriptions. These are summarized in Tables 2 and 3, respectively. For translation from a Web service to an agent service, WS2JADE handles these mismatches through its different components. Although the current version of WS2JADE is operational and offers many advantages over other tools, it can still be improved in a number of areas.

For example to increase the ontology reuse and avoid redundancy, the semantic mapping management component can be extended to detect semantic equivalence of two syntactically different generated concepts and keeps one of them only. In [2], the authors focus on this topic and outline some approaches. This version of WS2JADE has not yet implemented that specific feature. Another area is the interaction pattern translations. Web Service Choreography Description Language (WS-CDL) has been under development for same time. WS-CDL is considered as a layer above WSDL in the Web Service technology layer hierarchy. It describes a set of rules to explain how different partners may act in a conversation. In WS2JADE approach, we plan to convert BPEL4WS and WS-CDL (however not at this stage of WS-CDL development) into Agent Unified Modeling Language (AUML [10]) for the overall protocol representation in UML template. AUML is an extension of UML language for agents and has been used as a standard language to describe FIPA interaction protocols. The interaction translator and management will keep generated AUML documents in its protocol specification repository which can be looked up by client agents (or the client agents' designers) before using the service. In this version of WS2JADE we have not implemented the translation of WSCL to AUML. One reason is because the instability and immaturity in Web Service Choreography at this time evidenced by the suppression of WSCI (Web Service Choreography Interface) and WSCL (Web Service Choreography Language) by WS-CDL which is still in the first draft version.

Table 2. Communication Mismatch

FIPA agent communication	W3 Web Service communication
ACL/IIOP+HTTP	SOAP/HTTP
Asynchronous	Synchronous/Asynchornous
Stateful	Stateless

4   WS2JADE in Action

WS2JADE version 1.0, released in December 2004, implements the core components with a gray color in Figure 3. From a user perspective, WS2JADE provides two distinct tools: a tool for ontology generation and management and a tool for deploying and controlling Web Services as agent services. The ontology generation and management tool is offered through the combination of the ontology translator and the interaction pattern management components. The Web Service deploying and controlling tool is provided through combination of the service assignment mapping and the interaction pattern management components.

The ontology generation and management tool can be used alone from a command line with an input of a WSDL address and several user options. The options include whether the user want to generate ontologies for the immediate imported XML types which are defined in other documents, and where to store Web Service stubs and agent ontologies. Another option is to specify the name for the agent service. The tool has been developed under ciamas.wsjade.wsdl2jade package with Java main class ciamas.wsjade.wsdl2jade.WSDL2Jade.

In WS2JADE, the basic functionalities of a WSAG have been implemented in ciamas.wsjade.wsdl2jade.utils. WSAgent. The users can reuse the code or extend this base class to implement new functionalities, agent reasoning for example; since the base class supports message translation only. If the agent is designed

Table 3. Description Mismatch

FIPA agent service	W3 Web Service
Name - Name of the service	Names of services, port types, operations, etc.
Type - Type of the service	Type - Container of data type
Protocols - List of supported protocols	Choreography interfaces
Ontologies - List of supported ontologies and languages	XML Schemas - different namespaces imported
Ownership - The owner of the service	
	Binding - Protocol and data format specs for a port type
	Port - Single endpoint as combination of a binding and a port type

with some complex behaviors in which single threaded mode is preferred, the base class can then be modified so that the agent is single threaded without difficulties. WS2JADE's Web Service deploying and controlling tool allows Jade agents to deploy Web Services on the fly. The tool has been developed in *ciamas.wsjade.management* package with the main class for graphical administration interface at *ciamas.wsjade.management.utils.Admin*. WS2JADE operates as follows. First important parameters in the configuration panel need to be set. The gateway agent container is then started and new agents are created from the "Agent List" tab. After deploying the new agents, new WSAG agents appear on the Jade platform and ready to be assigned with Web Services. Once the Web Services are assigned they have to be activated in order to be used by client agents. The deployment process can be done simply as shown in Figure 5, in which "Amazon" and "Google" services are deployed with a few clicks and inputs of the real Amazon and Google Web Service addresses. The WSAG who provide these services are "AmazonProvider" and "GoogleProvider". Ontology packages are generated and compiled on the fly, and the Web Services are now available as services of these WSAG agents. The generated ontology package is stored under the "Jade output folder" and can be sent to a Web Server for downloading and using by the client agents. Also, since each WSAG is multithreaded, each of them can serve many Web Services at the same time. In the deployment process, these services must be assigned different agent-service names. These names must be different for one WSAG and may be the same with the real Web Service name. "WS List" tab shows a snapshot of all deployed Web Services and generated ontologies. To test the Web Service, the user may create a client agent to use the distributed ontologies to invoke the Web Service through WS2JADE. For Amazon and Google services in this example, simple client sample code can be found from the WS2JADE distribution. Figure 6 shows the output of running sample Amazon client agent in the top window and Google client agent in

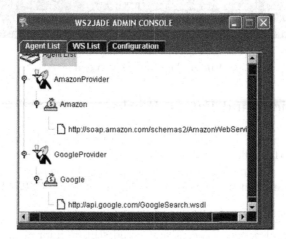

Fig. 5. Service Deployment Interface

the bottom window. In which, the Amazon client sends a request to search for all products with key word "Java" under "book" category from Amazon Web Service; the Google client agent finds five most popular links for a search on keyword "ciamas". Note that in order to compile and use these sample client agents, developer keys are required to access the Amazon and Google Web Services. These keys can be obtained for free after register from Amazon and Google Web Service websites [4,6].

WS2JADE version 1.0 is freely available for download from *http://www.it. swin.edu.au/centres/ciamas* . It should be noted that the current version WS2JADE does not cover all aspects described in part 3, and knowledge of the service URI that allows bypassing the discovery process is assumed . Implementation of these components is part of our on-going work.

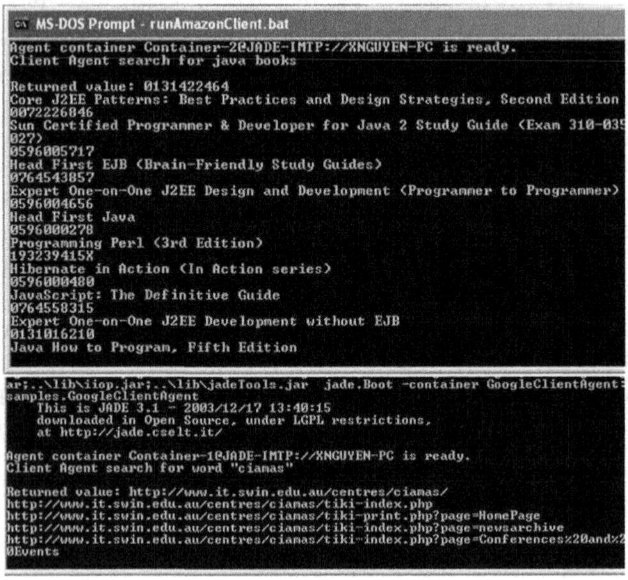

Fig. 6. Amazon client Agent

5   Conclusions and Future Works

The paper presents WS2JADE toolkit for integrating Web Services and Jade agent platform that allows Jade agents to offer Web Services as their own services at runtime. Comparison with another pioneer implementation from Sztaki is discussed to demonstrate the advanced features of the WS2JADE implementation.

Although WS2JADE offers many advantages over other existing tools its current version has also some aspects that could still be improved. One-way integration is one of them. At the moment we are reluctant in any implementations of the other end as we believe that, as discussed in the previous sections, there is still a lack of substantial theoretical work on the topic of agent to Web Service integration. This is a subject of our on-going research. As a contribution to Agent OpenNet, we are working on adding our WS2JADE deployment node into the agent network. Our current and future work also involves improvements of the semantic processing capability of the ontology management component. It also includes an implementation of auto-translation of Web Service choreography to agent protocol for dynamic mapping, invocation, and monitoring composite Web Services. This will support more dynamic and robust integration between the two platforms.

Acknowledgments

This work has been partially supported by the Adaptive Service Agreement and Process Management (ASAPM) in Services Grid project (AU-DEST-CG060081) and the EU FP6 Integrated Project on Adaptive Services Grid (EU-IST-004617). The ASAPM project is proudly supported by the Innovation Access Programme -International Science and Technology established under the Australian Government's innovation statement, Backing Australia's Ability.

References

1. *JADE (Java Agent Development Framework)*.
   http://sharon.cselt.it/projects/jade/.
2. *Web services Agent Integration Project*. http://wsai.sourceforge.net/index.html/.
3. *Agentcities Web services Working Group: Integrating Web services into Agentcities Technical Recommendation*. http://www.agentcities.org/rec/00006/, 2005.
4. *Amazon Web Service*. www.amazon.com/gp/aws/landing.html, 2005.
5. *FIPA: Interaction Protocol Specification*.
   http://www.fipa.org/repository/ips.php3, 2005.
6. *Google Web APIs*. http://www.google.com/apis/, 2005.
7. *uropean Coordination Action For Agent-Based Computing: AgentLink*.
   hhttp://www.agentlink.org/, 2005.
8. *W3C: Web Service Description Language (WSDL 1.1)*.
   http://www.w3.org/TR/wsdl, 2005.
9. *Web service Resource Framework*. http://www.golubs.org/wsrf, 2005.
10. B. Bauer, J. Muller, and J. Odell. Agent uml: formalism for specifying multiagent software systems, in agent-oriented software engineering. *Lecture Notes in Computer Science*, 1957:207–221, 2001.
11. D. Greenwood and M. Calisti. Engineering web service - agent integration. In *IEEE Systems, Cybernetics and Man Conference*, pages 1918 – 1925, the Hague, Netherlands, 2004. IEEE Computer Society.
12. L. Z. Varga and A. Hajnal. Engineering web service invocations from agent systems. *Lecture Notes in Computer Science*, 269:626–635, 2001.

Z-Based Agents for Service Oriented Computing

Ioan Alfred Letia, Anca Marginean, and Adrian Groza

Technical University of Cluj-Napoca
Department of Computer Science
Baritiu 28, RO-400391 Cluj-Napoca, Romania
letia@cs.utcluj.ro

Abstract. Ensuring reliability and adaptability of web services represents one of the main prerequisites for a larger acceptance of web services technology. We present an agent based framework to model the global behavior of atomic e-service and their composition using Z. We consider failures associated with web services and we try to handle runtime exceptions through formal methods for specification and verification of a composite service. In addition, our framework enforces the quality of services, in terms of answer time, by providing Z-agents responsible for these aspects.

1 Introduction

Service oriented computing [1] involves loosely coupled activities among two or more business partners. Orchestration describes the way the fine-grained services can interact with each other at the message level, to provide more coarse-grained business services which can be incorporated into workflows. The process oriented languages used in orchestration assume that the services combination is predefined, which deeply affects business process reconfiguration.

We handle the dynamic nature of service composition using a formal specification that integrates process oriented paradigms with ontological knowledge. Interleaving the execution phase with synthesis allows us to handle the predicted and unpredicted situations that may arise as a result of services enactments. On the one hand, the formal specification of the collaborating participants validates and guarantees the correct execution of the composite service. On the other hand, the transactional states within the composite service are flexibly manipulated in order to achieve fault tolerance and robustness [2] during service enactment.

Even though for simple service specification there is a formal approach included in WSDL version 2.0 [3], there has been essentially no formal work to understand the relationship between the global properties of the composite service and the local properties of its atomic components [4]. Out contribution consists in introducing a framework for modeling and specifying the global behavior of e-service composition through the Z language, in a multi-agent context.

2 An Agent Architecture for Handling Services

Effective composition of web services relies on concepts thoroughly studied in distributed computing, artificial intelligence and multiagent systems. Web services

J. Huang et al. (Eds.): SOCASE 2007, LNCS 4504, pp. 160–174, 2007.

are closely related to the agent programming paradigm. The definition of the web services architecture states that a web service is an abstract notion that must be implemented by a concrete agent [5]. Even so, web services do not currently assign any large role to agents and their interactions are still limited to simple request-response exchanges [6]. MAS mediating web services introduces a new kind of architecture, in which the communication patterns between agents representing a service can be considerably more varied and complex.

A MAS approach for web services enhances their capability of dealing with dynamic nature of environment and requirements. In our approach, the Z language is used by the agents for representing domain data, state model or current tasks. The Zeta-agent has the central role in the composition process, deciding the services that will be included and their enactment order. According to the Z-model, Zeta-agent recommends the next actions for each new state of the composed service and updates the current state based on responses of the enacted web services. If more than one service can be used for the current task, a reliable agent chooses the proper one as regard to some quality criteria. A type-checker agent decides how the operations are invoked.

2.1 Dealing with Failures

The composite service must deal with both deterministic and nondeterministic failures. A deterministic failure occurs when an atomic service returns a negative answer which makes the composite service to deviate from the initial plan. When the service does not respond within a time limit a nondeterministic failure has arisen, denoted by NF (Network Failure). Furthermore, handling failures must meet time constraints. The client provides a Δ_{max} time limit within which the composite task must be accomplished. This time is split according to some criteria and the milestones t_i are attached to each subtask. When a subtask is accomplished, the remaining milestones are adjusted. The reliability of the composed service is ensured by monitoring its state and re-synthesizing it when facing a failure.

The scenario used for testing our ideas is inspired from common medical activities. A physician initiates a consultation for a patient *Purgon* by sending the task *Consult(Purgon)* to a composition agent. Firstly, the generated composite service should check for the patient's insurance. It can exist more than one web services representing insurance houses, which may be simultaneously interrogated. If the patient has an insurance, the patient history service is asked to provide medical profile of the patient. Next, the physician consults the patient and he writes a prescription, which is sent to the the hospital secretary service. The secretary service updates the patient profile and it also requests the insurance house to pay for the consultation. Deterministic failures can appear on each state, for example if the patient does not have yet an insurance. The medical profile is not indispensable, therefore the consultation can begin after its allocated time has expired, even though the medical history service has not yet responded (nondeterministic failure).

3   Service Specification

This section covers the structural and behavioral specifications of the atomic services in Z language and it also identifies the knowledge involved in the composition process, encapsulated as business rules. *Operational reasoning*, including composition approaches in BPEL style, are fallible due to their limited adaptability restricted to some expected course of actions. In contrast, *the formal reasoning* views a composite service as a formula of which properties can be rigorously checked. The testing of composite service through execution, common to operational reasoning, is not sufficient to ensure its correctness in the presence of a large number of participating services and of the nondeterminism arising from the behavior of these services.

3.1   Structural Specification

The formal structural model is generated from the WSDL specification, translated by an automated parser to Z language [7] which defines a Z generic type for each WSDL component. The Z *Component* type represents the collection of all these generic types for operations, messages and their components (figure 1).

$$Component ::= element\langle\!\langle Element \rangle\!\rangle \mid part\langle\!\langle Part \rangle\!\rangle \mid message\langle\!\langle Message \rangle\!\rangle \mid$$
$$operation\langle\!\langle Operation \rangle\!\rangle \mid messageref\langle\!\langle MessageReference \rangle\!\rangle$$

```
┌── Operation ──────────────        ┌── MessageReference ──────────
│ Identifier                        │ Identifier
│ name : QName                      │ message : ℕ
│ type : OperationType              │ direction : Direction
│ messageRefs : seq ℕ               │
└────────────────────────           └──────────────────────────────
```

Fig. 1. Main types for WSDL description

The formal structural description makes possible the verification of referential integrity for the concrete service model. Instances of these generic types are used in type checking and behavioral specification of the service, and also for the effective enactment of the services.

3.2   Behavioral Specification

Inspired by flow composition, the behavioral specification augments the service specification with states modeling elements based on structural specification and additional knowledge contained by the WSDL-S descriptions. For each WSDL operation, a state transition is defined specifying its input values, the operation and the global task that can be accomplished. At this level, an operation op! is specified by its name and sequence of parameters corresponding to the elements of involved WSDL messages. This specification can be considered as a procedural translation of the structural one. For example, the transition *Pay*1 achieves the *pay* task by executing WSDL operation *Pay*1 with parameters *patient*? and *fee*?.

$$OPER ::= op1\langle\!\langle Operation1 \rangle\!\rangle$$

$$
\begin{array}{l}
\underline{\quad Pay1 \quad\rule{4.5cm}{0.4pt}} \\
fee? : \mathbb{N} \\
patient? : QName \\
task! : TASK \\
op! : OPER \\
\hline
op! = op1\langle\!\langle\, name == pay1, \\
\quad\quad param == \langle string\langle\!\langle\, value == patient?\,\rangle\!\rangle, \\
\quad\quad\quad\quad nat\langle\!\langle\, value == fee?\,\rangle\!\rangle\rangle\,\rangle\!\rangle \\
task! = pay \\
\end{array}
$$

$$
\begin{array}{l}
\underline{\quad Operation1 \quad\rule{3.5cm}{0.4pt}} \\
name : QName \\
param : \text{seq}\,ParamType \\
\end{array}
$$

There are no specified preconditions for transitions at this level. More behavioral information is added from the business rules at a centralized level. Inside a service community all the transitions with the same task as output are supposed to have the same final functionality from a service external viewpoint.

3.3 Business Rules

Current workflow technology is often too rigid [1], meaning that the agents have limited possibilities to reason on the information provided by the current WSDL specifications. Aiming to improve the adaptability and expressiveness, we consider the composed service as a business process and we propose the use of five different types of business rules together with the formal descriptions of services, both structural and behavioral.

Domain task coordination are structural rules that define a task. In order to have a shared representation of the tasks and their associated messages, we consider a domain ontology having the following structure

- concepts: *task, workflow task, messages*. If the composed service needs the human intervention, then the workflow task is used. There can be simple and composed tasks, for the latter being necessary to achieve some other tasks.
- object properties: *subtask, precedes: task → task*, defining execution constraints between tasks. The property *hasMessage: task → Message* allows defining the messages associated to a simple task.
- datatype properties: *typeOfTask* and *typeOfMessage*. We model two types of tasks: *typeT ∈ {informative, operational}* where *typeOfTask : task → typeT*. The operational tasks imply a change at the data level of the service (as in *updateHistory*), while the informative ones (*getHistory*) are used only to get some information thus being possible to be executed more than once, even in parallel when more than one service provides operations for them[1].

---

[1] A task without a *typeOfTask* property becomes operational if one of its subtasks is operational.

Regarding the type of message property *typeOfMessage* : *message* → *typeM*, there are two categories *typeM* ∈ {*input*, *output*}.

Querying the domain ontology provides for structural rules on tasks and execution constraints between them. Based on *precedes* and *substask* relations, one can identify different control patterns for subtasks[2]. Considering our scenario, the first two rules of the following schema represent a sequential pattern, whilst the last one is a parallel pattern. Having a hierarchical decomposition of tasks allows dealing with services of different granularity.

$$
\begin{array}{|l}
\hline
rules : \mathrm{seq}\,\mathbb{P}\,TASK \leftrightarrow TASK \\
\hline
rules = \{ \\
\quad \langle \{startConsult\}, \{endConsult\} \rangle \mapsto consult, \\
\quad \langle \{checkInsurance\}, \{getHistory\} \rangle \mapsto startConsult, \\
\quad \langle \{payConsultation, updateHistory\} \rangle \mapsto endConsult\} \\
\end{array}
$$

Operation constraints express at a global level the constraints of the task transitions through preconditions and effects. A task transition includes the task accomplished, the concrete operations from all known services that are able to achieve the task, preconditions, and effects expressed in terms of process and domain variables. *Process variables* are used to monitor the composition process, while the *domain variables* characterize the state of the composed service. The preconditions include both process and domain variables, whilst the effects are referring only to the domain variables.

In the example from figure 2, the *SimplePayGYes* transition accomplishes the *pay* task and it is possible only if the *pay* task has not been achieved yet (process variable condition) and the patient has insurance (domain variable condition). The effect will be the change of the domain variable *payConsultation*. The concrete operations are extracted from the behavioral specification of atomic services (in the below example, from the *Pay1* transition) and they are added to the sequence of operations in the *OpPayGYes* component of the *PayGYes* task transition. We have to observe that for the same task *pay* there can be more services providing achieving operations, and all these operations are included in *OpPayGYes*. The values of the process variables are not modified by transitions accomplishing a task as it can be seen in *HistoryComposedS* schema.

The task transition for the *pay* task is the result of merging these three schemes *PayGYes* == *OpPayGYes* ∧ *SimplePayGYes* ∧ *HistoryComposedS*. Due to the non-deterministic nature of the composite service, task transitions must catch all the possible situations. Therefore, more transitions are defined for the action of paying, one for the case the transaction was successfully done and one for the opposite case. When generating a plan, both of them are considered, but only one can be included: *PayG* == *PayGYes* ∨ *PayGNo*.

---

[2] These ones represent ontological control pattern. A different type of control patterns appears when the composite service handles preconditions and effects for each operation provided by the services.

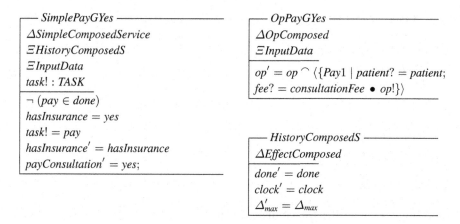

Fig. 2. *Pay* operation constraints

Message translation rules define the changes on the process and domain variables according to the response messages of the enacted services. Based on the message rules, at every step of the composition, a new state is computed.

Business entity constraints define input requirements for the evolution of variables through the entire composition process (the total cost of the composed service, time limit constraint Δ_{max}). In the formal specification they are expressed as axioms | *total_price* < 10 or as restrictions of type definitions. Therefore, when animating the Z formal model their truth values are verified in all transitions.

Time constraints are rules that manage instances of time allocated to each task and they influence the control flow of the composite service. Their role is to identify failures and to exclude the operations which have generated the nondeterministic ones from the re-planning process.

4   Z-Oriented Agents

In our framework we deal with a "community of web services" [8]. Each service is represented by an agent in the OAA community. The Z-based agents interact with Zeta tool[3] in order to define, verify, and animate the composite web service. Composition is done in a centralized manner, having the Zeta agent as the main orchestrator. The composed service is specified by the *ComposedService* type having as main components the process variables describing the computation process included in *EffectComposed* and *OpComposed* schemes, respectively the *InputData* and the *SimpleComposedService* for the domain specific variables.

4.1   General Algorithm for Composition

The composition process (see figure 3) is the result of interleaving planning with execution. It starts by generating plans according to available operations and

---

[3] http://uebb.cs.tu-berlin.de/zeta/

```
┌─ ComposedService ─────────────┐   ┌─ EffectComposed ──────────┐
│ InputData                     │   │ done : ℙ TASK             │
│ SimpleComposedService         │   │ clock : ℕ                 │
│ EffectComposed                │   │ Δmax : ℕ                  │
│ OpComposed                    │   ├───────────────────────────┤
└───────────────────────────────┘   │ clock ≤ Δmax              │
                                     └───────────────────────────┘
```

```
┌─ OpComposed ──────────────────┐
│ op : seq(ℙ OPER)              │
└───────────────────────────────┘
```

time constraints. Besides the information from the business rules, for each atomic operation x we consider the estimated time for execution T_e^x and the probability of execution P^x. We also use a factor β^x that indicates if the corresponding task for operation x is optional ($\beta^x = 0$) or obligatory ($\beta^x = 1$) and which is initially provided by the domain ontology. Then, the optional tasks may be adjusted ($0 \leq \beta^x \leq 1$) by the current client in order to indicate how much the task is desired. In this manner, client preferences are captured.

Each plan has a quality factor which depends on the estimated execution time T_e^x for each operation, on the probability of success P^x, and on the preferences given by the client to the optional tasks[4]. When there are more than one generated plans, the best one according to this quality factor is chosen. The chosen plan may not be entirely followed, after each enactment of a decided service, a plan being chosen from the new generated plans.

In case a new plan is picked having the estimated time $T^{newplan}$ the extra time is uniformly distributed to each operation. The service corresponding to the first operation is queried and its answer is waited. If the answer does not assure the achievement of the task, the available time of the operation is decreased and the probability of success for that operation is updated. Even if $T_{available}^x$ remains positive, another plan may become the best one for the moment according to the F_p factor. In case the same plan is chosen again, there is no need to redistribute the extra time. If no new plans can be generated and the requested *tasks* are not achieved, then the composition ends with failure.

4.2   Zeta Agent

This agent has access to the web services descriptions and to the business rules, acting as an orchestrator for the composite service.

Planning Phase. The process of generating the coordination artifact has two components that work together. A component that reasons above the domain task coordination rules and a lambda calculus component that works on states and tasks, trying to determine the sequence of actions related together by the states composition.

---

[4] For instance, the optional operations like *updateHistory* may have $\beta = 0.5$.

begin
  $currentplan = \{\}$
  repeat
    $\mathcal{P}$=compose($currentState, tasks, \Delta_{max}$)
    if $\mathcal{P} \neq \varnothing$ **then**
      for all $p \in \mathcal{P}$ **do**
        $F_p \leftarrow \sum_{k=1}^{N} \frac{\beta_x \cdot P^x}{T_e^x}, N =| p |$
      end for
      $newplan \leftarrow p$ with F_p maximal
      if $currentplan \neq newplan$ **then**
        $T_e^{newplan} \leftarrow \sum_{k=1}^{N} T_e^k, N =| newplan |$
        $T_s^{newplan} = \frac{\Delta_{max} - T_e^{newplan}}{N}$
        for all $x \in currentPlan$ **do**
          $T_{available}^x \leftarrow T_e^x + T_s^{newplan}$
        end for
      end if
      query the service associated to the first operation op of $newplan$
      $answers \leftarrow$ collectAnswers(op)
      if $updateState(currentState, answers, time) = \varnothing$ **then**
        $T_{available}^{op} \leftarrow T_{available}^{op} - time$
        $P^{op} \leftarrow update\_probability(T_{available}^{op})$
      else
        $currentState$=updateState($currentState, answers, time$)
      end if
    end if
  until $\mathcal{P} = \varnothing$ or achieved($currentState, tasks$)
end

Fig. 3. General algorithm for composition

$step : TASK \rightarrow (ComposedService \leftrightarrow ComposedService)$
────────────────────────────
$step = \lambda\, out : TASK \bullet$
  $\{\Delta ComposedService \mid UniformOps \bullet$
  $(\theta ComposedService, \theta ComposedService')\}$

$compose : \mathbb{P}\, ComposedService \rightarrow seq\, TASK \rightarrow \mathbb{P}\, ComposedService$
────────────────────────────
$compose = \lambda\, init : \mathbb{P}\, ComposedService \bullet \lambda\, tasks : seq\, TASK \bullet$
  if $\#tasks = 0$
  then $init$
  else
  if$((compose(init)(front(tasks)))) \cap (dom(step(last(tasks)))))) \subseteq \varnothing$
    then $\varnothing$
    else $step(last(tasks))(\!|compose(init)(front(tasks))|\!)$

The function $compose$ generates the state corresponding to the composite service in the planning phase. It receives the sequence of required tasks and the time constraints and determines the sequence of transactions that solve these tasks. It is a backwards recursive process that generates on each step a relation of possible *before* and *after* states for all the known transitions that accomplish the current task, through the *step* function.

When a task can not be reached from any state transition, the Z agent tries to decompose a task in subtasks and corresponding actions. A simple decomposition could be done as in *DecomposeTask* action. This action is one of the *UniformOps* that it is tried on every step of the composition.

$$UniformOps == UCheckInsurance \lor UGetHistory$$
$$\lor\ UPayConsultation \lor UUpdateHistory \lor UDecomposeTask$$

─── *DecomposeTask* ──────────────────────────────
$\Delta ComposedService$
$task? : TASK$
───
$\exists x : \text{seq } TASK \mid x \mapsto task? \in rules \land \neg\ compose(\{\theta ComposedService\})(x) \subseteq \varnothing \bullet$
    $\{\theta ComposedService'\} = compose(\{\theta ComposedService\})(x)$
───

The relation generated by the *step* function is used to reach the next state of the *ComposedService*. The second condition of compose function expresses the situation when none of the *before* state for the task matches the current state of the ComposedService. In these cases, the empty set is retrieved, meaning the composition of that task is not possible. There are three cases where the composition may fail in the planning phase: *(i)* when there is no available transition for the task, meaning there are no service accomplishing the task, *(ii)* there are no decomposition rules, and *(iii)* from one intermediate state it is not possible to do the transition for the task.

Updating Phase. The response messages of the enacted services are received by the Z agent and the new state of the composite service is computed according to message translation rules. An *UpdateState* transition must be able to update the composite service state according to the outcomes of the services inquired at the current step.

Correctness of Composition. All the states of the composed service are checked for validity according to the constraints expressed by domain task ontology and operation, respective message translation rules. The updating process of the service state following the receiving of a message is conditioned by the existence of a known transition possible in the current state that also follows the message translation rules. We check the acceptance of the transition given by the message by intersecting its *after* states set with all the possible *after* states from the current one through *stepEff* function. An unexpected message is considered to be valid for the composition process only if there is an available atomic transition taking the current state to the *after* update states.

$updateState == \lambda\, now : \mathbb{P}\, ComposedService;\ message : \mathbb{N};\ value : BOOLEAN;$
  $time : \mathbb{N};\ \bullet\, \{\Delta ComposedService\ |\ UniformUpdateState\, \bullet$
  $(\theta ComposedService, \theta ComposedService')\}\, (\!|now|\!)\, \cap\, stepEff\,(now)$

$stepEff : \mathbb{P}\, ComposedService \rightarrow \mathbb{P}\, ComposedService$

$stepEff = \lambda\, beforeState : \mathbb{P}\, ComposedService\, \bullet$
  $\{\Delta ComposedService\ |\ SimpleUniformOps\, \bullet$
  $(\theta ComposedService, \theta ComposedService')\}\, (\!|beforeState|\!)$

Interleaving planning with execution is the key element that assures the correctness of composition. If the service gets into an undesired state and compensation or rollback transitions are defined in the transition knowledge base, they will be included in the new plan, similar to the normal flow transitions.

The composite service (figure 5) is represented as a sequence of states, transitions being determined by querying a service *(op1?)* or receiving an answer *(mess1!)*. A sequence of a transition querying a service and one receiving the corresponding message determines the achievement of a task *(task1)*. The initial plan contains the request of *op2* that brought the composed service to the state 5 where *mess2* is the expected message. Due to the failure of receiving the expected message in the allocated time, a new plan is generated proposing *op3* as the first operation. The problem arises in state *7* where the composed service receives the expected message *mess3*, but also the message *mess2*. The wrong message is identified by the function *stepEff*, due to the fact that there is no defined atomic transition equivalent to one of those generated by the *updatingState* transition from state *7* and message *mess2*. The merging of states *6* and *8* is possible if their variables' values are consistent. In the case the merging is possible, there is no need to search back, it is enough to re-plan. In the contrary case, the state where the delayed message was expected must be identified and plans starting from both states *6* and *8* must be generated. The best one is followed.

__ *UpdateStateCheckInsurance* _____

$\Delta ComposedService$
$message? : \mathbb{N}$
$value? : BOOLEAN$
$clock? : \mathbb{N}$

$clock' = clock + time?$
$message? = checkInsuranceM$
$hasInsurance' = \mathbf{if}\ value? = yes\ \mathbf{then}\ yes\ \mathbf{else}\ no$
         $\mathbf{else}\ hasInsurance$
$payConsultation' = payConsultation;$
$op' = \varnothing$
$done' = done \cup (\mathbf{if}\ value? = yes\ \mathbf{then}\{checkInsurance\}\ \mathbf{else}\{\})$

Fig. 4. Update state schema for checkInsurance Message

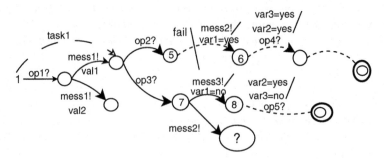

Fig. 5. A case of asynchronous message

The fulfillment of the requirements in each state constitutes another important component. The *achieved* function, receiving a service state and a task, checks if the specified *task* is accomplished in that state. A task is achieved if it is an atomic one, member of *done* variable, or there is a sequence of achieved subtasks that accomplishes the task. This function is used as a termination condition of the process.

$$
\begin{array}{l}
achieved : \mathbb{P}\, ComposedService \times TASK \rightarrow \mathbb{P}\, BOOLEAN \\
\hline
achieved = \lambda\, currentState : \mathbb{P}\, ComposedService;\ task : TASK\ \bullet \\
\quad \{x : ComposedService \mid x \in currentState\ \bullet \\
\quad\quad \textbf{if }\ task \in x.done\ \vee \\
\quad\quad\quad (\exists\, seqx : \mathrm{seq}\, TASK \mid seqx \mapsto task \in rules\ \bullet \\
\quad\quad\quad\quad \forall\, y : TASK \mid y \in \mathrm{ran}\, seqx\ \bullet\ yes \in achieved(currentState, y)) \\
\quad\quad \textbf{then }\ yes\ \textbf{else}\ no\}
\end{array}
$$

4.3   WSDL2Z Agent

The goal of this agent is to support the dynamic integration of web services. It defines a middle-ware trading service for retrieving service instances that match a given service specification in Z language. To obtain the WSDL specifications available in the community, the WSDL2Z agent broadcasts requests of type *getWSDL*. The obtained WSDL specifications are translated into Z and they are compared with the model encapsulated in the proper domain agent. If the specifications match, the names of the operations available in the community and which are considered useful for the current composition are provided to the Zeta agent.

4.4   Domain Agents

Domain agents provide a set of state variables and invariants for a specific domain. They have access to the domain task ontology and operation constraints rules. Identification of the domain task coordination rules from domain ontology is one of the responsibility of these agents.

For the medical domain, the *SimpleComposedService* type is defined by two domain variables *hasInsurance* and *payConsultation*, whilst the *InputData* type specifies four input variables of the composed service. Together with the types specific to the computation process *OpComposed* and *EffectComposed*, these two types define the ComposedService type (see section 3.3).

```
┌─ InputData ──────────────
│ patient : QName
│ consultationFee : ℕ
│ diagnosis : ℕ
│ receipt : QName
└──────────────────────────
```

```
┌─ SimpleComposedService ──────
│ hasInsurance : BOOLEAN
│ payConsultation : BOOLEAN
└──────────────────────────────
```

4.5 Reliability Agent

The specification in Z of non-functional properties of the composite service are verified by this agent. The reliability of a web service represents the probability that a request submitted to a service is correctly responded within the maximum expended time frame [9]. This agent has the task to compute the reliability value for each service from historical data about past invocations used when (i) the plans are generated, the zeta agent verifies or estimates if the plans can be executed within the time limit asked by the client and (ii) the OAA agent waits for answers only the time estimated by the reliability agent, after that it reports *NetworkFailure* and the zeta agent computes the next state.

4.6 Type-Checking Agent

Type checking is an important issue in web process composition. The process execution engines typically throw exceptions when they encounter incompatible types during data flow between activities. In order to use the output from one web service as input to another web service, it is often necessary to perform a data transformation. This agent acts as an intermediate layer between the client and the service. The agent converts the data type provided by the client and the data type supported by the service to Z representation and performs type checking. The agent ensures both that only compatible types are used while establishing a data flow, and also aids in decision making during the automated data flow.

5 Related Work

Automated composition of web services is an open research issue. On the one hand XML-based standards have been developed to formalize the composition of web services. This line is primarily syntactical and the interaction protocols are manually written. On the other hand, semantic approaches based on ontologies view the composition process as a goal oriented one, basically as a planning problem. Our approach starts from a formal specification is order to automate and validate the composition process, but also provides mechanisms to use onto-logical knowledge during the composition process. The main advantage relies on using the built-in state transition composition mechanism of the Z language. We advocate two strong points of the approach: i) it generates more reliable services, and ii) the composition expressivity in Z language is not limited to the reasoning capabilities of the description logic in OWL-S. Integration of agent technology with web services and semantic web is also aimed in [10] or [11]. In the for-mer, the integration of web services is coordinated with TucSon, while in latter OWL-S descriptions of web services together with production rules are used by workflow managers, developed in JADE, that build or complete the workflow responding to the user requests. In our approach, the composition process is not limited to the reasoning capabilities of the description logic.

A classical approach for monitoring the service execution consists in plan-ning as model checking [12]. Our zeta agent deals with such monitoring aspects, computing the current state using schema calculus. Specification of composite services also uses service chart diagrams [13], providing a good number of con-trol flow constructs. In our framework, the composition is verified at runtime, handling some unpredicted exceptions through re-planning.

Formal specification of complex systems uses a combination CSP-OZ-DC in [14] for the specification of processes, data and time. The WSAMI language was also proposed for the specification of a composite service [2]. Given the WSAMI specification of a service, an instance is automatically selected and com-posed upon a user request, according to the services that may be retrieved in the environment. In the web services context, the Z language was used for a formal specification of a constrained object model for the workflow composition [15]. The above research is focused on the design aspects, the resulted Z specification being compatible with UML. Our approach is more functional, the specification being animated by the Zeta tool and then executed by Z-based agents.

The value of flexible provisioning for service flows has been shown [16] by empirical evaluation in an experimental testbed.

6   Conclusions

In this paper, we introduce a framework for formal specification and verification of composite services. We considered failures associated with web services and we tried to handle such runtime exceptions by using formal methods. Using Z-based agents, a series of advantages exists: i) both process oriented knowledge and ontological knowledge are used in the composition process; ii) the operations of a service are represented by Z-schemes and the correctness of the composition is verified with the schema calculus; iii) the above mathematical framework is not limited to the reasoning capabilities of the description logic; iv) due to the existence of multi-agents, one can model more complex interactions between services, not only request-response messages;

We plan to enhance the animation capabilities of Zeta by introducing a tool for reasoning with the available Z specifications. Another challenge would be dealing with the preference concept for describing different importance levels for the composition rules. The framework is a step forward a functional system where the formal specifications and the semantic descriptions could work together for improving the collaboration between services in open environments.

Acknowledgments

We are grateful to the anonymous reviewers for useful comments. Part of this work was supported by the grant 27702-990 from the National Research Council of the Romanian Ministry for Education and Research.

References

1. Singh, M.P., Huhns, M.N.: Service-Oriented Computing:Semantics, Processes, Agents. John Wiley and Sons, Chichester West Sussex (2005)
2. Issarny, V., Sacchetti, D., Tartanoglu, F., Sailhan, F., Schibout, R., Levy, N., Talamona, A.: Developing ambient intelligent systems: A solution based on Web Services. Automated Software Engineering **12** (2005) 101–137
3. W3C: Web Services Description Language (WSDL) version 2.0 part 1: Core language. Technical report, W3C, available at http://dev.w3.org/cvsweb/ checkout /2002/ws/desc/wsdl20/ (21 February 2005)
4. Bultan, T., Fu, X., Hull, R., Su., J.: Conversation specification: A new approach to design and analysis of e-service composition. In: 12th International World Wide Web Conference (WWW'2003), Budapest, Hungary (2003) 403–410
5. Booth, D., Haas, H., McCabe, F., Newcomer, E., Champion, M., Ferris, C., Orchard, D.: Web services architecture. Technical report, W3C, available at http://www.w3.org/TR/2003/WD-ws-arch-20030808/ (8 August 2003)
6. Paurobally, S., Jennings, N.R.: Protocol engineering for web services conversations. Int J. Engineering Applications of Artificial Intelligence **18** (2005) 237–254
7. Jacky, J.: The way of Z - Practical Programming with Formal Methods. Cambridge University Press, Cambridge (1998)
8. Berardi, D., Calvanese, D., De Giacomo, G., Lenzerini, M., Mecella, M.: Automatic composition of e-services that export their behavior. In: International Conference on Service Oriented Computing, Trento, Italy (2003)
9. Zeng, L., Benatallah, B., Dumas, M., Kalagnanam, J., Sheng, Q.Z.: Quality driven web service composition. In: 12th International World Wide Web Conference (WWW'2003), Budapest, Hungary (2003) 411–421
10. Morini, S., Ricci, A., Viroli, M.: Integrating a MAS coordination infrastructure with web services. In: Workshop on Web-Services and Agent-based Engineering at AAMAS, New York, NY, USA (2004)
11. Negri, A., Poggi, A., Tomaiuolo, M., Turci, P.: Agents for e-Business Applications. In: 5th International Joint Conference on Autonomous Agents and Multiagent Systems, Hakodate, Japan, ACM Press (2006) 907–914
12. Pistore, M., Barbon, F., Bertoli, P., Shaparau, D., Traverso, P.: Planning and monitoring web service composition. In: ICAPS04, Workshop on Planning and Scheduling for web and grid Services, Whistler, Canada (2004)
13. Maamar, Z., Benatallah, B., Mansoor, W.: Service chart diagrams - description application. In: 12th International World Wide Web Conference (WWW'2003), Budapest, Hungary (2003)

14. Hoenicke, J., Olderog, E.R.: Combining specification techniques for processes, data and time. In Butler, M., Petre, L., Sere, K., eds.: Integrated Formal Methods. LNCS 2335. Springer-Verlag (2002) 245–266
15. Albert, P., Henocque, L., Kleiner, M.: A constrained object model for configuration based workflow composition. In: Business Process Management Workshops. (2005) 102–115
16. Stein, S., Jennings, N.R., Payne, T.R.: Flexible provisioning of service workflows. In: 17th European Conference on Artificial Intelligence. (2006)

Author Index

Lecture Notes in Computer Science

For information about Vols. 1–4376

please contact your bookseller or Springer

Vol. 4426: Z.-H. Zhou, H. Li, Q. Yang (Eds.), Advances in Knowledge Discovery and Data Mining. XXV, 1161 pages. 2007. (Sublibrary LNAI).

Vol. 4425: G. Amati, C. Carpineto, G. Romano (Eds.), Advances in Information Retrieval. XIX, 759 pages. 2007.

Vol. 4424: O. Grumberg, M. Huth (Eds.), Tools and Algorithms for the Construction and Analysis of Systems. XX, 738 pages. 2007.

Vol. 4423: H. Seidl (Ed.), Foundations of Software Science and Computational Structures. XVI, 379 pages. 2007.

Vol. 4422: M.B. Dwyer, A. Lopes (Eds.), Fundamental Approaches to Software Engineering. XV, 440 pages. 2007.

Vol. 4421: R. De Nicola (Ed.), Programming Languages and Systems. XVII, 538 pages. 2007.

Vol. 4420: S. Krishnamurthi, M. Odersky (Eds.), Compiler Construction. XIV, 233 pages. 2007.

Vol. 4419: P.C. Diniz, E. Marques, K. Bertels, M.M. Fernandes, J.M.P. Cardoso (Eds.), Reconfigurable Computing: Architectures, Tools and Applications. XIV, 391 pages. 2007.

Vol. 4418: A. Gagalowicz, W. Philips (Eds.), Computer Vision/Computer Graphics Collaboration Techniques. XV, 620 pages. 2007.

Vol. 4416: A. Bemporad, A. Bicchi, G. Buttazzo (Eds.), Hybrid Systems: Computation and Control. XVII, 797 pages. 2007.

Vol. 4415: P. Lukowicz, L. Thiele, G. Tröster (Eds.), Architecture of Computing Systems - ARCS 2007. X, 297 pages. 2007.

Vol. 4414: S. Hochreiter, R. Wagner (Eds.), Bioinformatics Research and Development. XVI, 482 pages. 2007. (Sublibrary LNBI).

Vol. 4412: F. Stajano, H.J. Kim, J.-S. Chae, S.-D. Kim (Eds.), Ubiquitous Convergence Technology. XI, 302 pages. 2007.

Vol. 4411: R.H. Bordini, M. Dastani, J. Dix, A.E.F. Seghrouchni (Eds.), Programming Multi-Agent Systems. XIV, 249 pages. 2007. (Sublibrary LNAI).

Vol. 4410: A. Branco (Ed.), Anaphora: Analysis, Algorithms and Applications. X, 191 pages. 2007. (Sublibrary LNAI).

Vol. 4409: J.L. Fiadeiro, P.-Y. Schobbens (Eds.), Recent Trends in Algebraic Development Techniques. VII, 171 pages. 2007.

Vol. 4407: G. Puebla (Ed.), Logic-Based Program Synthesis and Transformation. VIII, 237 pages. 2007.

Vol. 4406: W. De Meuter (Ed.), Advances in Smalltalk. VII, 157 pages. 2007.

Vol. 4405: L. Padgham, F. Zambonelli (Eds.), Agent-Oriented Software Engineering VII. XII, 225 pages. 2007.

Vol. 4403: S. Obayashi, K. Deb, C. Poloni, T. Hiroyasu, T. Murata (Eds.), Evolutionary Multi-Criterion Optimization. XIX, 954 pages. 2007.

Vol. 4401: N. Guelfi, D. Buchs (Eds.), Rapid Integration of Software Engineering Techniques. IX, 177 pages. 2007.

Vol. 4400: J.F. Peters, A. Skowron, V.W. Marek, E. Orłowska, R. Słowiński, W. Ziarko (Eds.), Transactions on Rough Sets VII, Part II. X, 381 pages. 2007.

Vol. 4399: T. Kovacs, X. Llorà, K. Takadama, P.L. Lanzi, W. Stolzmann, S.W. Wilson (Eds.), Learning Classifier Systems. XII, 345 pages. 2007. (Sublibrary LNAI).

Vol. 4398: S. Marchand-Maillet, E. Bruno, A. Nürnberger, M. Detyniecki (Eds.), Adaptive Multimedia Retrieval: User, Context, and Feedback. XI, 269 pages. 2007.

Vol. 4397: C. Stephanidis, M. Pieper (Eds.), Universal Access in Ambient Intelligence Environments. XV, 467 pages. 2007.

Vol. 4396: J. García-Vidal, L. Cerdà-Alabern (Eds.), Wireless Systems and Mobility in Next Generation Internet. IX, 271 pages. 2007.

Vol. 4395: M. Daydé, J.M.L.M. Palma, Á.L.G.A. Coutinho, E. Pacitti, J.C. Lopes (Eds.), High Performance Computing for Computational Science - VECPAR 2006. XXIV, 721 pages. 2007.

Vol. 4394: A. Gelbukh (Ed.), Computational Linguistics and Intelligent Text Processing. XVI, 648 pages. 2007.

Vol. 4393: W. Thomas, P. Weil (Eds.), STACS 2007. XVIII, 708 pages. 2007.

Vol. 4392: S.P. Vadhan (Ed.), Theory of Cryptography. XI, 595 pages. 2007.

Vol. 4391: Y. Stylianou, M. Faundez-Zanuy, A. Esposito (Eds.), Progress in Nonlinear Speech Processing. XII, 269 pages. 2007.

Vol. 4390: S.O. Kuznetsov, S. Schmidt (Eds.), Formal Concept Analysis. X, 329 pages. 2007. (Sublibrary LNAI).

Vol. 4389: D. Weyns, H.V.D. Parunak, F. Michel (Eds.), Environments for Multi-Agent Systems III. X, 273 pages. 2007. (Sublibrary LNAI).

Vol. 4385: K. Coninx, K. Luyten, K.A. Schneider (Eds.), Task Models and Diagrams for Users Interface Design. XI, 355 pages. 2007.

Vol. 4384: T. Washio, K. Satoh, H. Takeda, A. Inokuchi (Eds.), New Frontiers in Artificial Intelligence. IX, 401 pages. 2007. (Sublibrary LNAI).

Vol. 4383: E. Bin, A. Ziv, S. Ur (Eds.), Hardware and Software, Verification and Testing. XII, 235 pages. 2007.

Vol. 4381: J. Akiyama, W.Y.C. Chen, M. Kano, X. Li, Q. Yu (Eds.), Discrete Geometry, Combinatorics and Graph Theory. XI, 289 pages. 2007.

Vol. 4380: S. Spaccapietra, P. Atzeni, F. Fages, M.-S. Hacid, M. Kifer, J. Mylopoulos, B. Pernici, P. Shvaiko, J. Trujillo, I. Zaihrayeu (Eds.), Journal on Data Semantics VIII. XV, 219 pages. 2007.

Vol. 4379: M. Südholt, C. Consel (Eds.), Object-Oriented Technology. VIII, 157 pages. 2007.

Vol. 4378: I. Virbitskaite, A. Voronkov (Eds.), Perspectives of Systems Informatics. XIV, 496 pages. 2007.

Vol. 4377: M. Abe (Ed.), Topics in Cryptology – CT-RSA 2007. XI, 403 pages. 2006.